Programming
the Pico

Programming the Pico

Learn Coding and Electronics with the Raspberry Pi Pico

Simon Monk

Copyright © 2022 by Simon Monk. All rights reserved.
Published by MonkMakes Press.

No part of this book may be reproduced or transmitted in any form or by any means, electronic or mechanical, including photocopying, recording, or by any other information storage and retrieval system, without written permission from the publisher.

Limit of Liability/Disclaimer of Warranty: While the publisher and author have used their best efforts in preparing this book, they make no representations or warranties with respect to the accuracy or completeness of the contents of this book and specifically disclaim any implied warranties of merchantability or fitness for a particular purpose. No warranty may be created or extended by sales representatives or written sales materials. The advice and strategies contained herein may not be suitable for your situation. You should consult with a professional where appropriate. Neither the publisher nor author shall be liable for any loss of profit or any other commercial damages, including but not limited to special, incidental, consequential, or other damages.

Technical Editors: Ian Huntley and Mike Bassett
Composition: Patricia Wallenburg (TypeWriting)

ISBN 979-8-464-88217-1

First Printing

10 9 8 7 6 5 4 3 2 1

Contact us by email at support@monkmakes.com.

You can find out more about MonkMakes at monkmakes.com and about Simon's other titles at simonmonk.org.

Contents

Preface ix
Acknowledgments xiii

1   Introduction 1
   Microcontrollers 1
   A Tour of the Pico 3
   Input Output Pins 4
   Third-Party Boards 7
   Programming 7
   Summary 8

2   Getting Started 9
   Installing the Thonny Python Editor 9
   Getting Familiar with Thonny 11
   Connecting the Pico to your Computer 14
   Interacting with Your Pico 15
   Summary 17

3   MicroPython Basics 19
   Numbers 19
   Variables 20
   Strings 22

	Programs	22
	Looping Forever and Blinking	25
	Example Code	26
	Blinking SOS	27
	For Loops	29
	If and else	31
	Summary	33
4	**Functions**	**35**
	What Are Functions?	35
	Parameters	36
	Return Values	39
	Random Numbers	40
	Named Parameters	41
	Summary	42
5	**Lists and Dictionaries**	**45**
	Lists	45
	List Example	48
	Strings as Lists	48
	Dictionaries	50
	A Morse Code Translator	52
	Summary	58
6	**Modules, Classes and Files**	**59**
	Importing from Modules	59
	Documentation	61
	Useful Built-in Modules	61
	Classes	63
	String Methods	66
	Files and Exceptions	68
	Exceptions	70

Contents

	File Counter Example	71
	Summary	72
7	**Inputs and Outputs**	**73**
	Digital Outputs	73
	Digital Inputs	74
	Analog Outputs	77
	Analog Inputs	80
	Summary	83
8	**Electronics**	**85**
	Solderless Breadboard	86
	Soldering Pins onto Your Pico	88
	Components	90
	Making LEDs Blink	91
	RGB LEDs	94
	Servomotors	97
	Summary	102
9	**Advanced Inputs and Outputs**	**103**
	Interrupts	103
	Timer Interrupts	105
	High Speed PWM	107
	Multicore Support	108
	Programmable IO	109
	Summary	110
10	**Sensors**	**111**
	Variable Resistors	111
	Temperature	114
	Light	115
	Summary	117

11	**Displays** 119
	OLED Displays. 119
	An OLED Clock 124
	NeoPixel Displays 125
	Summary 130
12	**Other Programming Languages.** 131
	CircuitPython. 131
	Arduino C/C++. 137
	Summary 139
	Appendix 141
	Raspberry Pi Pico Pinout 141
	Index 143

Preface

The Raspberry Pi Foundation has achieved enormous success with its range of single board computers. Since the release of the original Raspberry Pi in 2014, the Raspberry Pi has evolved into a machine that makes a perfectly respectable replacement for a more traditional desktop computer. Added to that, the Raspberry Pi versions Zero to 4 and now the Pi 400 have always provided access to GPIO pins to allow sensors, displays and all sorts of electronics to be interfaced directly with the Raspberry Pi. The Raspberry Pi Pico is a radical departure from all previous versions of the Raspberry Pi. Unlike other Raspberry Pis, the Pico has no operating system and no built-in interfaces for keyboard, mouse and monitor—rather, the Pico is aimed squarely at electronics enthusiasts and educators wishing to learn about physical computing.

The Raspberry Pi Pico (let's just call it a Pico) is not a regular computer, but rather a microcontroller. That is, it is not intended for general purpose computing but rather it is designed to help you make electronic projects—to, if you like, be the brain of such projects. For example, a Pico might be pressed into service to make a robot, controlling motors and a loudspeaker to make sounds, or it might be used to display temperature or other sensor readings on a small LCD screen.

The gold standard microcontroller board, widely used in education and by hobbyists, is the Arduino Uno. The Pico is firmly in the same territory as the Uno, but is a much more up-to-date and powerful design. What's more, it is priced extremely competitively, even finding its way onto the cover of Hackspace magazine as a giveaway. While an Arduino is programmed in the lean and efficient programming language C++ (which you can also use on the Pico), the Pico's dual-core ARM processor is capable of running the more memory-hungry Python language, that is the most popular programming language in the world and widely used by educators. Most people find it easier to get started programming with Python than they do C++, and the official implementation of Python recommended by the Raspberry Pi Foundation (MicroPython) is perfectly fast enough for most projects.

This book teaches you Python at the same time as learning how to make use of the Pico. No prior programming or electronics knowledge is assumed or required to learn Python and get your Pico doing some pretty cool stuff.

When it comes to exploring the hardware side of things, you are going to need a side-order of electronic bits and bobs to make the most of your Pico. Buying the parts you need can be tricky if you are new to electronics and so, in this book, we use the Monk Makes Breadboard Kit for Raspberry Pi Pico. This kit was designed specifically for this book and includes a good range of basic components to get you started.

If you prefer not to get too involved in the electronics, then you can just use the Pico as a vehicle for learning Python using the Pico's built-in LED and a small length of wire will be enough to try out a whole range of projects.

HOW TO USE THIS BOOK

Learning to program, like any skill, requires practice and simple straight-forward examples that you can follow. In this book, you will be led step-by-step through example programs, intermingled with explanations and background.

The first chapter is an introduction to the Pico, providing a guided tour of the board and explaining a bit more about its features and just what you can use it for. This should leave you itching to use your Pico for real, and so Chapter 2 is a guide to getting started—you will upload your first program onto your Pico, and carry out a few experiments, just to get used to using the device.

Preface

Chapters 3 to 6 are all about Python, and are illustrated with an extensive Morse code example using the Pico's built-in LED to flash out messages. By using the built-in LED, there is no need to use any other components, and the example gradually gets more sophisticated as your Python knowledge grows.

Chapters 7 to 10 show you how to use MicroPython to use electronics sensors and buttons and how to interface with servomotors.

This book is centred around the officially recommended MicroPython; however, MicroPython has a competitor in the shape of Adafruit's CircuitPython. In Chapter 12, you will learn about the differences between MicroPython and CircuitPython, as well as learning where to go for more CircuitPython resources. Having used this book to learn Python, you will not find it hard should you wish to try the CircuitPython alternative. This chapter also provides a brief introduction to programming your Pico in C++, using the Arduino software, should you want to try it out.

EXAMPLE CODE

All the code examples for this project are available for download from GitHub at https://github.com/simonmonk/prog_pico_ed1. The simplest way to get the example files onto your computer is to download the ZIP file at https://github.com/simonmonk/prog_pico_ed1/archive/main.zip and then extract the contents into a convenient location.

ELECTRONICS HARDWARE

You can find information on the MonkMakes Breadboard Kit for Raspberry Pi Pico at: https://www.monkmakes.com/pico_kit1.

Acknowledgments

Many thanks to Ian, Mike and Patty for their help turning my comma-abusing manuscript into something more closely resembling a book.

Thanks also, of course, to my wife Linda, for putting up with my periodic need to write books.

The breadboard layouts for this book were created using the excellent Fritzing Software (fritzing.org). Other diagrams were drawn using Inkscape (inkscape.org).

1
Introduction

In this chapter, you will learn about the Pico's place in the world of electronics, its features and what you can use it for.

The Pico's processor chip (the RP2040) is also finding its way onto a number of other educational and hobbyist boards, so in learning how to use the Pico you will also be learning how to use these boards.

MICROCONTROLLERS

The term *microcontroller* really refers to the chip at the centre of the Pico—most of the rest of the Pico is concerned with providing power, and making the connections to that chip accessible. So, it seems reasonable to refer to both the chip and the whole circuit board as a *microcontroller*.

A microcontroller is a computer, in the sense that it has a processor and memory and runs a program that does things—perhaps read a sensor, do some calculations and then show information on a display, or switch something on and off.

One of the differences between a microcontroller and a single board computer is that a microcontroller can usually only run one program at a time. Thus a second (*normal*) computer is required to write programs and upload them onto the microcontroller to be run, whereas a single board computer (like the Raspberry Pi 4) has its own operating system and does not need a second computer to program it. Figure 1-1 shows how a microcontroller is typically used.

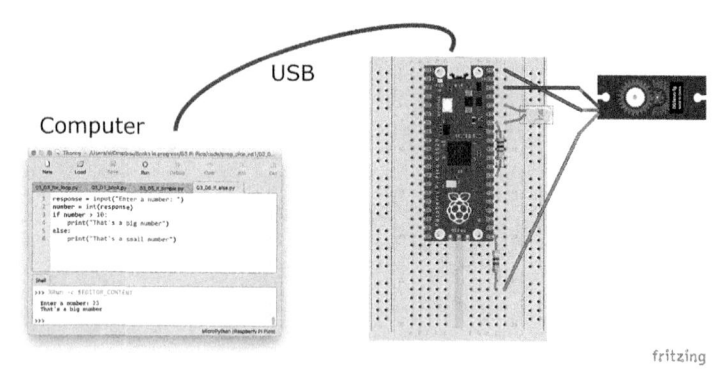

FIGURE 1-1 Using a Microcontroller.

    This process of writing code for a microcontroller using a second computer and code-editing software and then *embedding* the programmed Pico as an independent *brain* or *controller* for the project is why the whole field of microcontrollers is often referred to as *embedded computing*. The example shown in Figure 1-1 is for a light-meter project. Once the project is complete, it can be detached from your main computer and work independently. If course, if your goal is just to learn programming or electronics, you may just choose to take it all apart and make the next project.

Introduction

A TOUR OF THE PICO

Figure 1-2 shows the Pico, highlighting items of interest on the board.

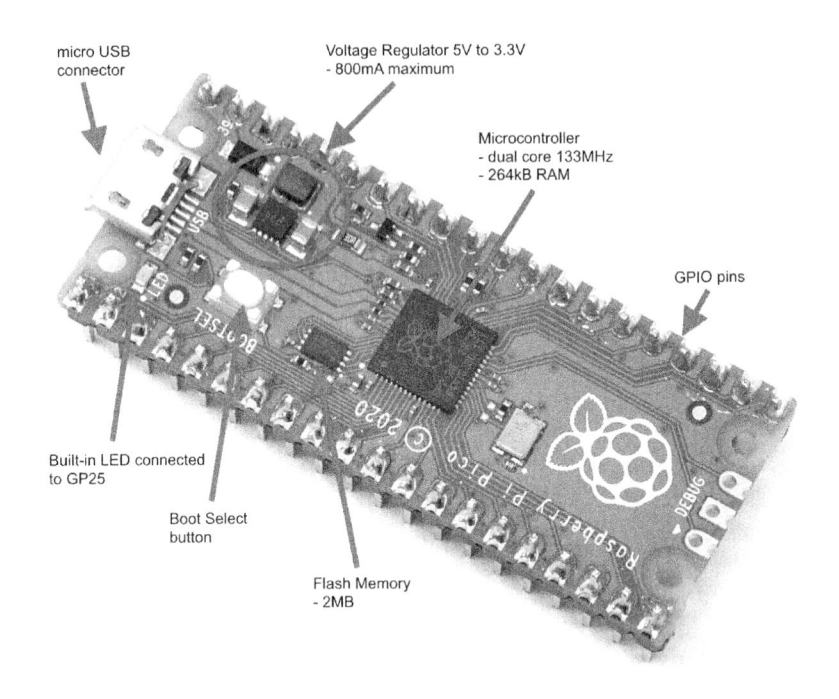

FIGURE 1-2 The Raspberry Pi Pico.

The micro USB connector is used to attach your Pico to your computer. This serves the dual purpose of supplying power to the Pico and acting as a data interface between your computer and the Pico, so that you can transfer programs onto the Pico and interact with it in other ways.

Moving clockwise around the board, you can see an area marked as the *Voltage Regulator*. This takes the 5V supplied by the USB connector and converts it to the 3.3V used by the Pico's microcontroller. This 3.3V supply is also available for you to use as a power source in your projects, and can supply up to 800mA.

The IC (integrated circuit) in the center of the board is the actual microprocessor. This is a processor (the RP2040) designed by the Raspberry Pi organisation that has two processing cores and plenty of memory, yet can operate in a low current mode if required, making it very suitable for battery-powered projects.

The GPIO (General Purpose Input Output) pins along the two long sides of the Pico allow you to attach components like LEDs, servomotors, displays, and sensors to the Pico.

Continuing around the board, we have a separate 2MB flash memory chip. 2MB may not sound like very much, but is actually a lot more than most microcontrollers have, and does open up the possibility of storing low resolution images or sound files to use in your projects.

The Pico has one built-in LED that you can control from your programs.

Finally, the *Boot Select* button is rarely used but, if pressed while the Pico powers up, it puts the Pico into a special mode that allows you to reinstall MicroPython, or indeed instal other firmware (such as Adafruit's CircuitPython that you will meet in Chapter 12).

Unusually for a microcontroller board, the Pico has no reset button. Some people don't like this, but unplugging it and plugging it back in again will cause a reset. If you do need to reset your Pico in this way, I recommend pulling the larger full-size end of the USB lead out of your computer, because this is a lot more robust than the micro USB connector on the Pico.

INPUT OUTPUT PINS

We will not get to use the GPIO (General Purpose Input Output) pins until later chapters, but if you are familiar with other microcontrollers and are interested in the electronics side of things then prepare to be wowed by the features available on the Pico.

Figure 1-3 shows a map of the GPIO pins of the Pico and, as you can see, there are multiple features available on many of the pins. Apart from the power and control pins, all the rest of the pins can be used as digital inputs or outputs. They can also all be used to generate PWM (Pulse Width Modulation) signals for controlling the brightness of LEDs, or providing control pulses to servomotors.

Introduction

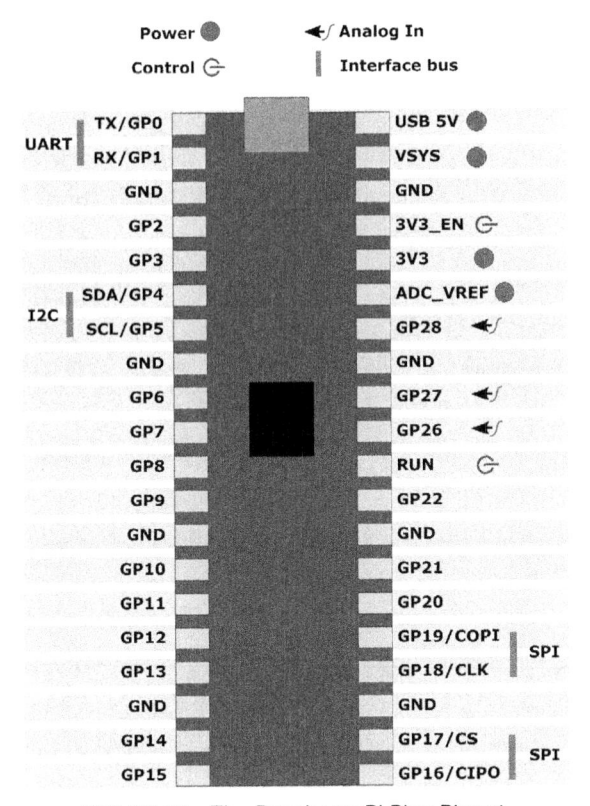

FIGURE 1-3 The Raspberry Pi Pico Pinout.

Just three of the pins can be used as analog inputs. That makes them suitable for attaching analog sensors for example to measure temperature or light.

Power and Control Pins

The Pico has a flexible and well designed power supply. When connected to USB, the USB 5V pin comes directly from the USB of the computer or power supply you are connected to. This can be useful if you want to attach devices that require 5V to your Pico. However, USB is not the only option for powering the Pico—once you are finished programming it and are ready to power it from a battery pack, you can use any voltage supply of between 1.8V and 5.5V connected to VSYS. Even if this voltage is under 3.3V, the buck-boost voltage regulator of the Pico will generate a reliable 3.3V for the Pico to use. This 3.3V supply is also available for use in your electronics projects on the pin labelled 3V3 (a short-hand for 3.3V).

The 3V3_EN pin has the strange function of turning off the voltage regulator (and so the whole Pico) when connected to GND. This can be useful, in specialised very low current projects, and allows an external circuit to put the Pico into standby mode.

The RUN pin is somewhat confusingly named, because it's really a reset pin that, when connected to GND, will reset the Pico.

Data Buses

Although you can turn LEDs on and off, use push switches and so on with normal digital inputs and outputs, many peripherals that you may want to connect to your Pico such as displays, accelerometers, and other advanced sensors use a bus interface—a standard protocol for making the connections. As is often the case with standards, there are several different standards of buses that you can use, and the Pico handles any that you are likely to want.

The first of these, UART (Universal Asynchronous Receiver Transmitter), isn't really strictly speaking a bus, but it is a fairly common way of connecting peripherals such as GPS receivers, Smartcard and barcode readers. By default UART can be used on pins GP0 and GP1. However, there is a second UART that you can optionally use, and associate with pins GP4 and GP5 or GP8 and GP9.

Similarly, the very common I2C bus is available on pins GP4 and GP5, with a second bus optionally available on various other pin combinations.

A single SPI bus which is commonly used by some small LCD and OLED displays is available on GP16, 17, 18 and 19.

The important point about all these advanced options is that they only tie-up a pin if you decide to use them. If not, then these pins can just be used as digital inputs and outputs. This flexibility of pin use means that you should be able to use your Pico to connect to a wide range of external devices at the same time, satisfying the needs of the most complex of projects.

THIRD-PARTY BOARDS

The RP2040 chip at the centre of the Pico is also available for other manufacturers to use in Pico-like boards such as the Pimoroni Tiny 2040 and the Arduino Nano RP2040 Connect. These boards can be programmed in the same way as a Pico, with some differences for the number of pins or built-in peripherals (Figure 1-4).

FIGURE 1-4 Pico-compatible boards from Pimoroni and Arduino.

While its great to have a choice, its probably best to start with a regular Raspberry Pi Pico.

PROGRAMMING

Programming, or coding if you prefer, is the process of writing a program that tells a computer, or in this case the Pico's microcontroller, what to do. You can think of a program as a list of instructions to be carried out. For example, to make an LED blink on and off, the instructions might (in English) be as follows. (If it helps, imagine someone being the processor and standing by a light switch with one hand on the switch and the other holding a watch—your job is to provide instructions that are clear and unambiguous.)

1. Turn the LED on
2. Wait for half a second
3. Turn the LED off
4. Wait for another half second
5. Go back to step 1

Following these instructions, you can see that, as well as simply performing commands like turning the LED on and off, we also need control command like step 5 that allow commands to be repeated.

When we write a program for the Pico, we type the text for the program into an editor using a programming language. In this case, the programming language is a version of the Python programming language called MicroPython. Having written the program, we save it as a file. If we put this file onto the Pico, then the Pico will run it for us.

SUMMARY

Now that we have found out a bit more about the Raspberry Pi Pico and some of its features, we can build on this in the next chapter, installing the software that we need onto our computer and running our first program on the Pico.

2
Getting Started

One of the great things about the Raspberry Pi Pico is that all you need to get started is a USB lead to connect it to your computer, and an editor in which to write your Python code before transferring it to the Pico. The Pico even has a built-in LED that we can control from your program code.

In this chapter you will learn how to install the software you need, how to interact over the USB connection with the Pico and how to transfer a program over to it. We choose to use the Thonny editor, since this is the one recommended by the Raspberry Pi Foundation, but several alternatives are available.

INSTALLING THE THONNY PYTHON EDITOR

To install programs onto your Pico, you will need a second computer. This can be a Windows, Mac or Linux computer, or even, a Raspberry Pi such as the Raspberry Pi 4 or 400. The Thonny Python editor is available for all these platforms, and actually comes pre-installed on Raspberry Pi OS.

The installation process is slightly different for each platform. You will find details here on the Thonny website: https://thonny.org/.

Windows

From the Thonny home page, click on the *Windows* link in the *downloads* section of the Thonny homepage. This will download an *exe* installer. Once it has downloaded, run the installer (Figure 2-1). You can accept all the defaults that the installer offers, clicking Next to take you from step to step.

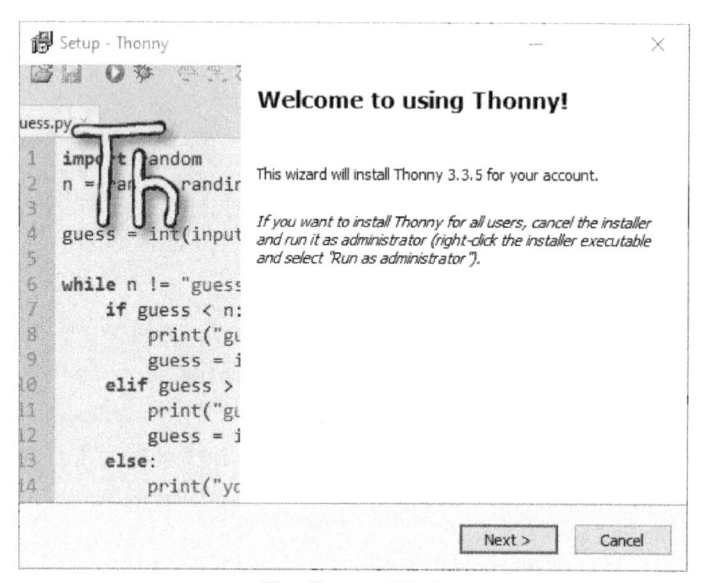

FIGURE 2-1 The Thonny Windows Installer.

When the installer has finished, click on the *Finished* button and you should now see Thonny on your *Start* menu.

Mac OS

On Mac OS, Thonny is installed using a package installer. Click on the Mac download link and, when the .pkg files has finished downloading, open it (Figure 2-2).

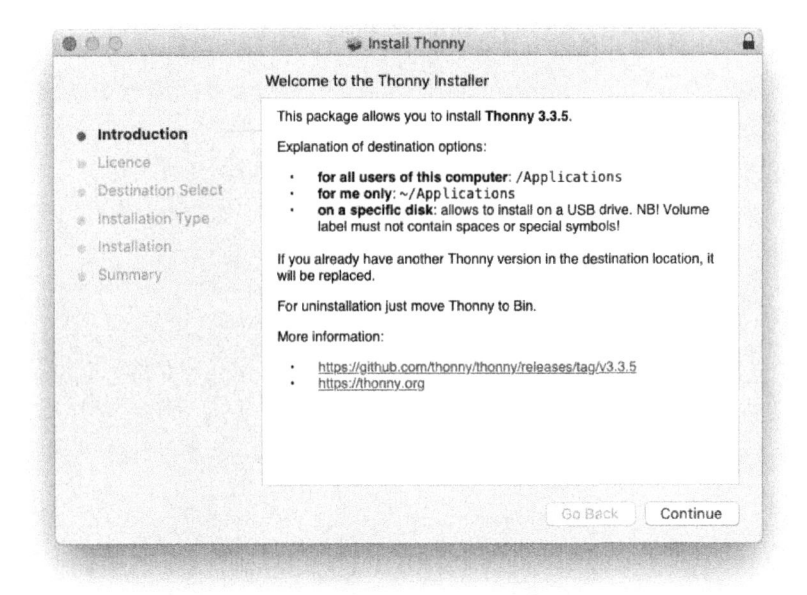

FIGURE 2-2 The Thonny Mac OS Package Installer.

Click on *Continue* after each step of the installation until installation is complete.

Linux

Installing Thonny on a Linux distribution is best done using the pip3 Python Package Installer by running the command

```
$ pip3 install thonny
```

If your distribution uses the apt package manager, you can install using

```
$ sudo apt install thonny
```

GETTING FAMILIAR WITH THONNY

The Thonny editor is designed with beginners in mind, so you can run it in different modes. In *simple* mode, much of the complexity of the software and less-frequently-used menu options are hidden. You can change the mode of Thonny at any time between *simple, regular* and *expert,* as your knowledge and confidence grow.

Figure 2-3 shows the Thonny window in regular mode on Windows. This is the mode it appears in when it first starts.

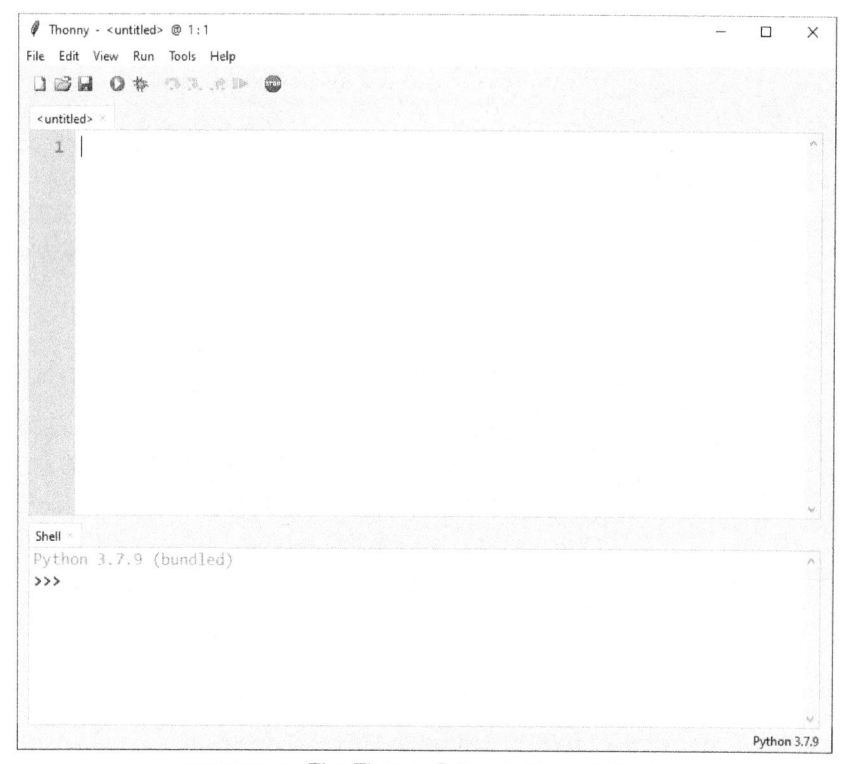

FIGURE 2-3 The Thonny Editor in Normal Mode.

This is a little unfortunate if you are a beginner as, to change the mode to simple, you will need to open the *Options* panel from the *Tools* menu and change the UI (User Interface) mode to simple (as shown in Figure 2-4).

For the change in UI mode to take effect, you have to quit Thonny and then restart it. When it restarts, it should look like Figure 2-5.

There are three main areas to the Thonny editor, the toolbar at the top, the editor area in the middle and the area labeled *Shell* at the bottom.

Programs are contained in text files, and the toolbar area provides access to buttons that allow you to create new program files, as well as loading and saving files from your file system. Python programs all have the extension of .py.

Getting Started

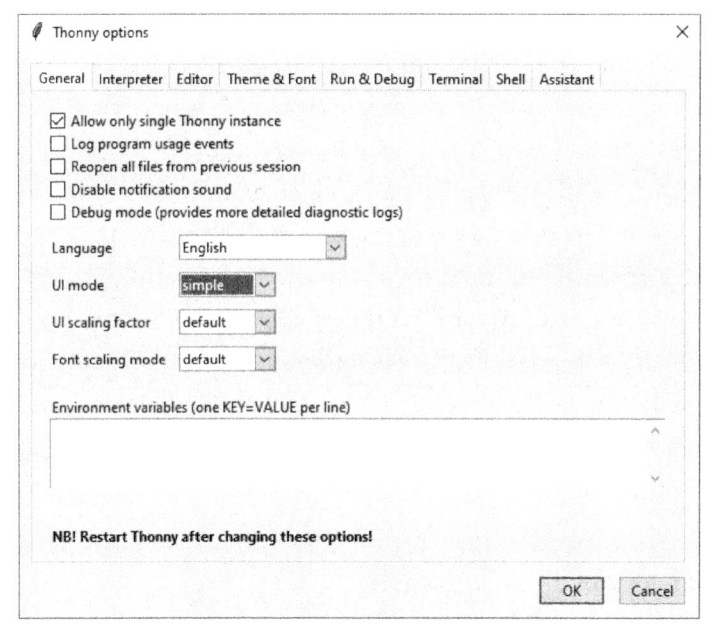

FIGURE 2-4 Changing the UI Mode.

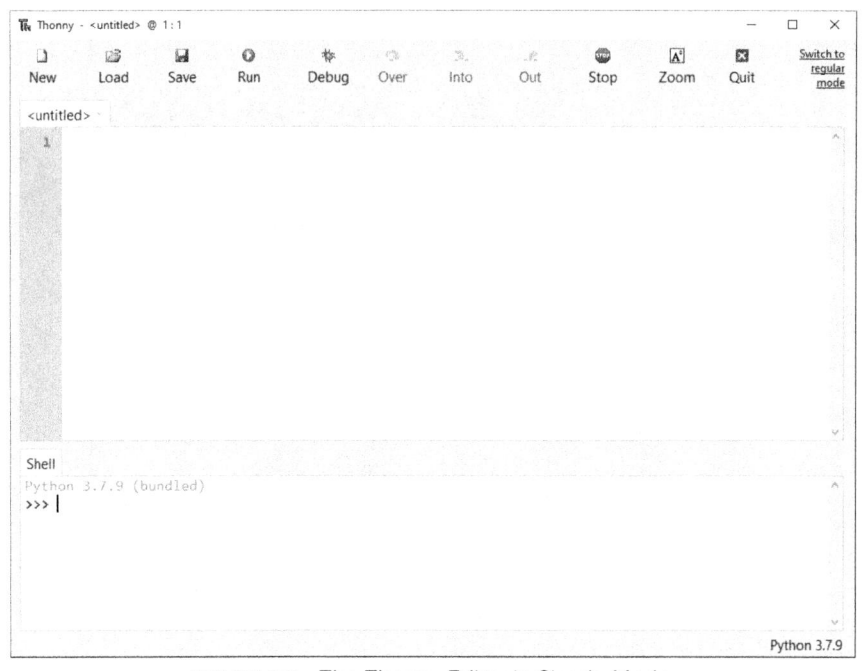

FIGURE 2-5 The Thonny Editor in Simple Mode.

As well as buttons for managing the program files, you will also find buttons in the toolbar for running the program being edited, for stopping the running program, and also for using the debugger that allows you to diagnose problems with your program by stepping through it one step at a time.

The editor area allows you to modify the program code. You will notice that there are line numbers in grey at the side of the editor. This is useful as, when there are problems running the code, the description of the problem will refer to a line number, helping you to find the source of the problem.

The area of Thonny labelled *Shell* allows you to run individual Python commands directly on the Pico, without having to write a whole program. This is a great way to experiment with Python and to try things out before committing them to a program file.

CONNECTING THE PICO TO YOUR COMPUTER

The Raspberry Pi Pico has a micro USB connector on one end, and this needs to be connected to your computer using a micro USB lead (Figure 2-6).

FIGURE 2-6 Connecting the Pico to a computer.

Any USB lead that you have lying around is likely to work just fine. The only exception to this is that some (not many) USB leads save a few cents by not including the data wires. These *power only* USB leads will provide power to your Pico, but will not allow you to transfer programs to it. So, if it seems like you are doing everything right but you still can't transfer a program onto the Pico, try swapping out the USB lead.

The Pico does not have a power LED—the LED it does have is under programming control so, when it is plugged in, do not expect the LED to light.

INTERACTING WITH YOUR PICO

Now that we have connected the Pico to the computer, Thonny has access to it so we can start experimenting with the Pico using Thonny's Shell feature.

Lets start by checking that our Pico is listening to us by typing **2 + 2** in the Shell area and pressing the Return key, as shown in Figure 2-7.

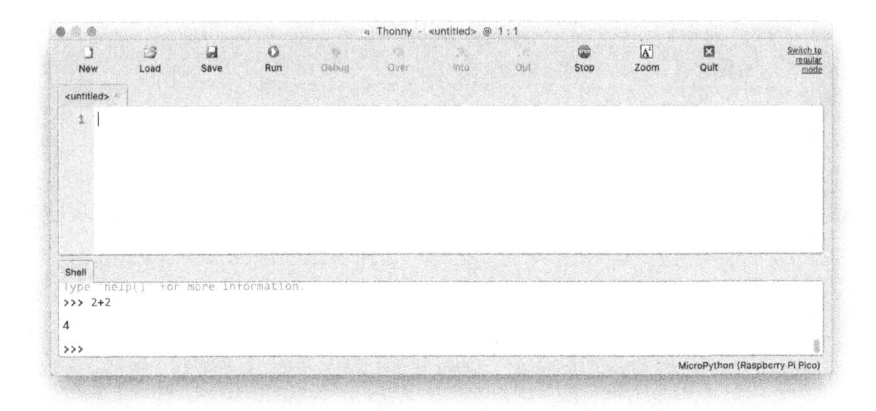

FIGURE 2-7 Experimenting in the Shell.

Any text, such as **2 + 2**, that you type into the Shell will be sent to the Pico to evaluate. In this case, we are asking the Pico to do a sum for us and display the result. We will use the Shell a lot as a way of experimenting with Python.

Programming the Pico

Let's take this a step further by sending the commands to the Pico to turn on its built-in LED. So, now type the following into the Shell, a line at a time. Python, like most programming languages is case sensitive. That is, if the code says Pin you must enter Pin and not pin or PIN.

```
>>> from machine import Pin
>>> led = Pin(25, Pin.OUT)
>>> led.on()
```

Note that I have shown the >>> prompt that indicates that the Pico is ready and listening, you don't need to type that.

Once you have typed the led.on() command, the small green LED on the Pico next to the USB connector should light. You can turn the LED off again with the command

```
>>> led.off()
```

You might want to try turning the LED on and off a few times by repeating just the led.on() and led.off() commands (Figure 2-8).

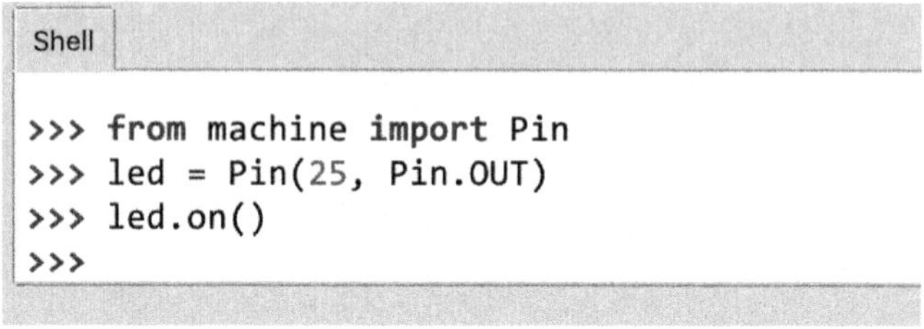

```
Shell
>>> from machine import Pin
>>> led = Pin(25, Pin.OUT)
>>> led.on()
>>>
```

FIGURE 2-8 Turning the LED on from the Shell.

To save having to type the whole commands each time, use the up arrow key on your keyboard. This will recall the last line you typed, and is an extremely useful feature of the editor.

16

SUMMARY

In this chapter, you have learnt how to install Thonny so that you can write programs for and interact with your Pico. You have experimented a little with the Shell area of Thonny to do sums and turn the built-in LED on and off. In the next chapter you will carry out some more experiments with the Shell and also upload your first program onto the Pico.

3
MicroPython Basics

In this chapter, we will explore some of the basics of Python using our Pico, getting used to some key ideas like numbers, variables and ways of making our code take different paths. We will also use the Pico's built-in LED as a first step towards a project that we will gradually develop to use the LED to flash out Morse Code messages.

NUMBERS

Let's start by carrying out a few more experiments using Thonny's Shell feature. Start Thonny, and connect your Pico to your computer. Start by repeating an experiment from Chapter 2 and type **2 + 2** after the >>> prompt.

There are two important types of numbers in Python: integers (or ints for short) which are whole numbers like 1, 2, 3 etc. and floating point (*floats*) that have a decimal place such as 1.1, 2.5, 10.5 etc. In many situations, you can use them interchangeably. Now try typing the following after the prompt

```
10 + 5.5
```

As you can see, the result is 15.5 as you would expect. Try out a few sums for yourself. If you want to do multiplication use * and for division use /.

Note that from now on things you need to type will be proceeded by >>> and the Pico's response will start on a new line.

Enter the following into the Shell.

```
>>> from random import randint
>>> randint(1, 6)
5
```

You are likely to get a different answer from 5. Repeat the `randint(1, 6)` line a few times—by pressing the up-arrow key on your keyboard to recall your last command, and you will see a succession of random numbers between 1 and 6. You have made a dice of sorts!

The first line you typed imports the `randint` function from the module `random`. Modules are used to contain Python code that you may not need all the time. This helps to keep the size of the code small enough to run on a Pico.

VARIABLES

Try typing the following line into the Shell. This will assign the value 10 to the variable x

```
>>> x = 10
```

You can put spaces either side of the = sign or not, it's a matter of personal preference. I think it looks neater with spaces, so that's the standard I will be sticking to in this book.

This line of code is using a *variable* called x. Now that the variable x has been given (or assigned) the value 10, try just typing x in the Shell

```
>>> x
10
```

Python is telling us that it remembers that the value of x is 10. You don't have to use single letter names for variables, you can use any word that starts with a letter and you can include numbers and the underscore (_) character. Sometimes you need a variable name that's made up of more than one human language word. For example, you might want to use the name *my number* rather than x. You can't use spaces in variable names and so you use the underscore character to join the words like this: `my_number`.

By convention, variables usually start with a lowercase letter. If you see variables that start with an uppercase letter, it usually means that they are what are called *constants*. That is, they are variables whose value is not expected to change during the running of the program, but that you might want to change before you upload your program onto your Pico.

Returning to our experiments in the Shell, now that Python knows about x you can use it in sums. Try out the examples below

```
>>> x + 10
20
>>> x = x + 1
>>> x
11
```

In this last command (x = x + 1) we have added 1 to x (making 11) and then assigned the result to x, so that x is now 11. Increasing a variable by a certain amount is such a common operation that there is a shorthand way of doing it. Using += combines the addition and assigning a value into one operation. Try the following

```
>>> x += 10
>>> x
21
```

You can also use round brackets (and) to group together parts of an arithmetic expression. For example, when converting a temperature from degrees Centigrade to degrees Fahrenheit, an approximation is to multiply the temperature in degrees C by 5/9 and then add 32. Assuming that the variable c contains a temperature in degrees C you could write the following in Python

```
>>> c = 20
>>> f = (c * 9/5) + 32
>>> f
68.0
```

The (and) are not strictly necessary here, because MicroPython will automatically perform multiplication and division before it does addition, but it can make the code easier to follow.

STRINGS

Computers are really good at numbers, that is after all what they were originally created for. But, as well as doing math, computers often need to be able to use text. Most often, this is to be able to display messages that we can read. In computer-speak bits of text are called *strings*. You can give a variable a string value, just like numbers. Type the following into the Shell.

```
>>> s = "Hello"
>>> s
'Hello'
```

So first we assign a value of *Hello* to the variable s. The quotation marks around *Hello* tell Python that this is a string and not say the name of some variable. You can use either single or double quotation marks, but they must match. We then check that s does contain *Hello*.

Rather like adding numbers, you can also join strings together (called concatenation). Try the following in the Shell

```
>>> s1 = "Hello"
>>> s2 = "World"
>>> s = s1 + s2
>>> s
'HelloWorld'
```

This might not be quite what we were expecting. Our strings have been joined together, but it would be better with a space between the words. Lets fix this

```
>>> s = s1 + " " + s2
>>> s
'Hello World'
```

PROGRAMS

The commands that we have been typing into the Shell are single-line commands that just do one thing. You can see how typing them one after the other

MicroPython Basics

leads to them being run or *executed* one after the other. For example, revisiting this example

```
>>> s1 = "Hello"
>>> s2 = "World"
>>> s = s1 + s2
>>> s
```

The four command lines are executed one after the other as fast as we can type them. If you put all four lines into a file, and then tell the Pico to run the whole file, you will have written a program.

Let's try that now—but first we need to make one slight change. When you use the Shell and simply type the name of a variable, the Shell will display the value in that variable. However, when running the same commands as a program, you must use the `print` function for any value that you want to display. So, click on *New* in Thonny to create a new program, and then type the following lines into the editor window (Figure 3-1).

```
s1 = "Hello"
s2 = "World"
s = s1 + s2
print(s)
```

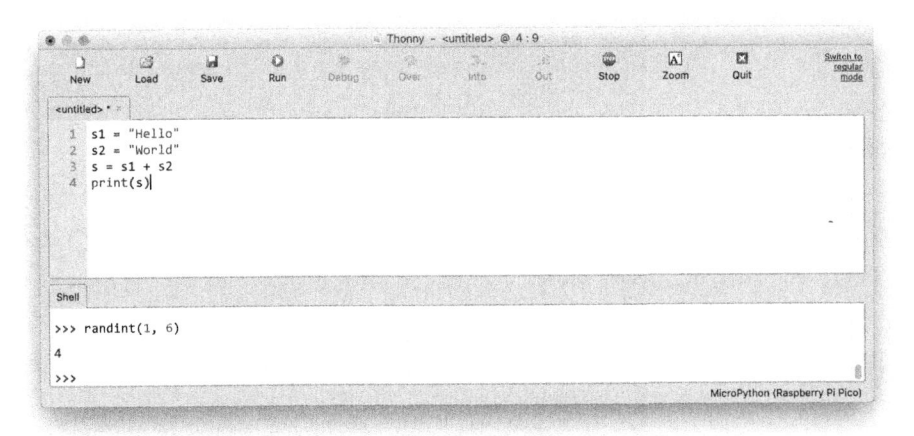

FIGURE 3-1 Hello World in Thonny.

23

Notice how the program is labelled *<untitled>* at the top, because we have not yet saved the file anywhere. That is why, when you click on the *Run* button, Thonny won't run the program immediately, but instead asks you to save the file, either somewhere on your computer or on the Pico itself (Figure 3-2).

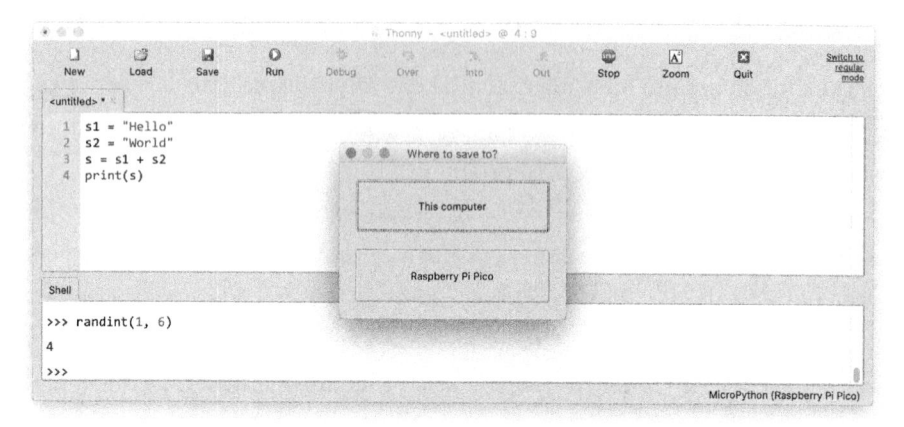

FIGURE 3-2 Saving to your computer vs saving to the Pico.

Choosing the *Save to your computer* option will open a file dialog window, where you will select a location on your computer to save the file. If you select save to the Pico, then the file will be saved to the Pico's really simple built-in file system. The choice as to where to save the program file has some important implications. Saving to the Pico itself has the benefit that, once saved to the Pico, the Pico no longer needs Thonny or your computer to be able to run the program. Also, if you name the file *main.py*, it will automatically run that program every time it is powered up.

Saving on the Pico does run the risk that should something happen to your Pico, your program could be lost. So, when we are still experimenting with small programs, its fine to save it on the Pico, but if you were writing a longer program, that you wanted to keep safe, you would save it on your computer first and then (when ready to deploy it properly onto the Pico) you would save a copy of the program onto the Pico.

For now, lets' save it on the Pico—so select the option *Raspberry Pi Pico* and then give the file the name *main.py* as shown on Figure 3-3.

MicroPython Basics

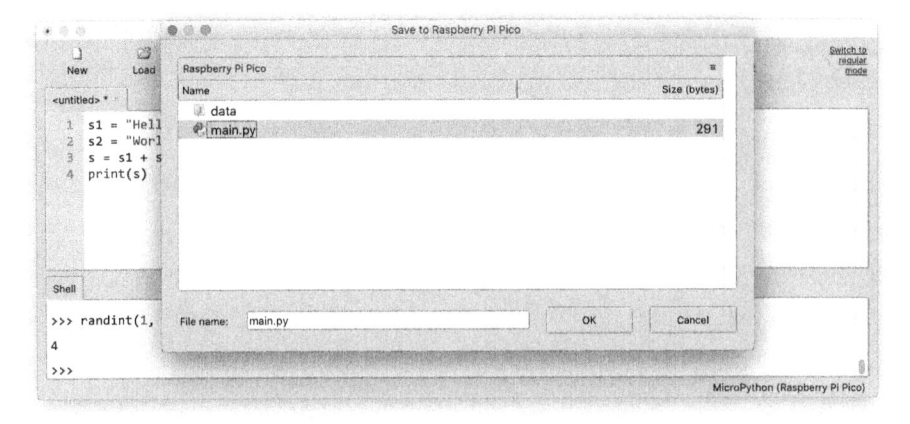

FIGURE 3-3 Saving a program onto the Pico.

Now that the file has been saved, click on the Run button and you will see the message *helloworld* appear in the Shell—the Shell is now just the area where program output appears.

LOOPING FOREVER AND BLINKING

Generally speaking, you don't want the program running on a Pico to end—it's not like running an app on your computer that you can just quit when you have finished using it. A program on a Pico will basically run until you unplug it. This is why you find something called a *while loop* at the end of most Pico programs. As it's a loop in the code that continues forever, let's call it the *eternal loop* from now on.

Type the following program into the editor area of Thonny, replacing the code that was there.

```
from machine import Pin
from utime import sleep

led = Pin(25, Pin.OUT)

while True:
    led.on()
```

```
sleep(0.5)
led.off()
sleep(0.5)
```

Before we look at what this code does, or actually run it, an important point about indentation needs to be made. You will notice that after the `while True:` line, the remaining lines are all indented by four spaces. This indicates that those lines belong *inside* the `while` command. Python insists that they line up exactly (or you will get an error message) and the convention is to use four spaces of indentation. Helpfully, in Thonny, when you press the *TAB* key you get four spaces inserted for you.

Save and Run the program under the name main.py and you should see the built-in LED start to blink. Just to prove that the code is indeed now installed on the Pico, unplug the Pico, quit Thonny and then plug your Pico back in. Now, the only thing that your computer is doing for the Pico is supplying it with power over USB. Your Pico should automatically run the program and start blinking.

Here's how the code works. The first two lines, import `Pin` and `sleep` into our program so that we can use them. This is necessary because much of MicroPython, especially the parts specific to the Pico are kept in code modules. You will learn more about this in Chapter 6.

The variable `led` is then associated to pin GP25, the internal pin of the Pico's microcontroller that is connected to the built-in LED.

The command `while True:` tells the micro:bit to repeat all the indented lines (after the : on the end of the line) forever. The first of these indented lines turns the LED on, the next delays for 0.5 of a second, then turns the LED off, followed by another delay after which the whole cycle repeats itself indefinitely.

EXAMPLE CODE

Its a good idea to type a few programs into Thonny while you are learning. However, all the code examples for this project are available for download from GitHub at https://github.com/simonmonk/prog_pico_ed1, so its a good idea to download them, especially as the code examples start to get a bit longer.

MicroPython Basics

To download the files, visit the link above in your browser and then click on the Code button and then Download ZIP (Figure 3-4), When the ZIP archive has downloaded, extract its contents to get the Python files contained in a folder called prog_pico_ed1-main.

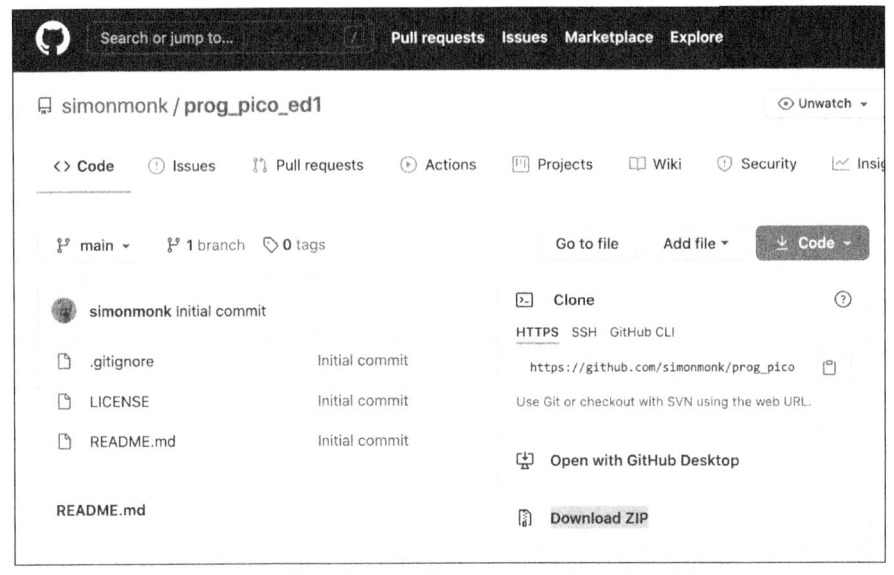

FIGURE 3-4 Downloading the Example Code from Github.

The blink program that you just wrote is in the code download, the file is called 03_01_blink.py. Each program file is prefixed by the chapter number (in this case 03). Try loading this file in Thonny by clicking on the *Open* button and navigating to wherever you extracted the ZIP file to.

BLINKING SOS

Morse Code is an ancient type of signalling code, that uses a system of short and long pulses, either of sound (long and short beeps) or flashes of light. We are going to use the Pico's built-in LED to signal for us. In Morse code, short pulses (of sound or light) are called *dots* and long pluses are called *dashes*.

The starting point for our Morse code adventure is to blink out the emergency signal S.O.S. (Save Our Souls) which comprises three short pulses (*dots*) for the letter S, followed by three long pulses (*dashes*) for the letter O, followed by three dots for the second S, and then repeating. Strictly speaking, the SOS should be like a word, with a gap before the next sequence of ...---... but the convention for SOS in an emergency setting is that it is just an unbroken repetition of the letters S and O: '...---...---...---...' and so on.

We can modify our blink program so that it generates this SOS signal by adding some more `led.on()`, `led.off()` and `sleep` commands to what is already in the eternal loop.

```
from machine import Pin
from utime import sleep

led = Pin(25, Pin.OUT)

while True:
    # S
    led.on()
    sleep(0.2)
    led.off()
    sleep(0.2)
    led.on()
    sleep(0.2)
    led.off()
    sleep(0.2)
    led.on()
    sleep(0.2)
    led.off()
    sleep(0.6)
    # O
    led.on()
    sleep(0.6)
    led.off()
    sleep(0.6)
    led.on()
    sleep(0.6)
    led.off()
    sleep(0.6)
    led.on()
```

```
sleep(0.6)
led.off()
sleep(0.6)
```

Try running the program 03_02_sos.py. You should see the ever repeating pattern of three dots followed by three dashes. Note the use of the # comment mark to tell us that the code that follows is for the letter S or O. Any text after # is not program code, but rather comments that help the programmer and others looking at the code understand what is going on.

The code is now starting to get a little long and repetitive. In the next section we will use a `for` loop to improve this a little, and in the next chapter we will streamline this code even more by use of *functions*.

FOR LOOPS

In this section you will learn about `for` loops. This means telling Python to do things a number of times rather than just once. This is similar to the eternal `while` loop, except that this is not eternal—we will specify before it starts how many times it should repeat before the program carries on and does something else (or, if there is nothing else to do, just quits).

To get a feel for how this works, type the following code into a new editor window in Thonny, or load and run the example 03_03_for_loop.py.

```
for x in range(1, 10):
    print(x)
```

When you run the program, you should see the following output in the Shell. You may need to scroll up to see the start of the output.

```
1
2
3
4
5
6
7
8
```

```
9
>>>
```

This program has printed out the numbers between 1 and 9 rather than 1 and 10 as you might be expecting. The range command has an exclusive end point—that it, it doesn't include the last number in the range, but it does include the first.

There is some punctuation here, that needs a little explaining. The round brackets are used to contain what are called parameters. In this case, range has two parameters: the start of the range (1) and the end (10) separated by a comma.

The for..in command has two parts. After the word for there must be a variable name. This variable will be assigned a new value each time around the loop. So, the first time it will be 1 the next time 2 and so on. After the word in, Python expects to see a list of things. In this case, that list of things is a sequence of the numbers between 1 and 9.

The print command also takes an argument which displays it in the Python Shell. Each time around the loop, the next value of x will be printed out.

If we wanted to, we could use a for loop to change our SOS blinking program to repeat the dot and dash blinking. You will find this example in 03_04_sos_for.py. It should work exactly the same as 03_02_sos.py, but the code is a bit shorter.

```
from machine import Pin
from utime import sleep

led = Pin(25, Pin.OUT)

while True:
    # This loop does 3 dots (S)
    for x in range(1, 4):
        led.on()
        sleep(0.2)
        led.off()
        sleep(0.2)
    sleep(0.4) # Gap between S and O
    # This loop does 3 dashes (O)
```

```
for x in range(1, 4):
    led.on()
    sleep(0.6)
    led.off()
    sleep(0.6)
```

The first thing to note here is that even through we want three blinks, the range has to be 1 to 4, rather than 1 to 3 because, as we mentioned before, the end point of the range is exclusive. It's easiest just to remember to set the end of the range to be one more than the number of times you want something to happen.

The other point of interest in the code is that an extra sleep of 0.4 is needed between the dots flashing and the dashes flashing, because this extra delay can nolonger be built in.

IF AND ELSE

Most programs are more than just a simple list of commands to be run over and over again. They can make decisions, do some commands only if certain circumstances apply. They do this using a special Python word called `if` and its optional counterpart `else`.

Let's illustrate the use of the `if` command by making the program ask for you to type a number into the Shell and then say something about that number. Type the following code, into a new editor tab, or load the program 03_05_if_simple.py.

```
response = input("Enter a number: ")
number = int(response)
if number > 10:
    print("That's a big number")
```

Run the program and, in response to the question, enter a number greater than 10—you will get a response in the Shell saying that the number is big. Run the program again, and this time enter a number less than or equal to 10—you should not see a message this time.

Note the use of the input command, that provides a prompt in the Shell for you to enter a number, and whatever you enter will be assigned to the variable response. Whatever you type, even if its a number, will be a string and so must be converted to an integer using the int command before we can make comparisons with other numbers (in this case, the number 10).

The remainder of the line after the word if is called a condition. That means the condition that must be true for the code inside the if (indented) to run. Conditions can use

- < (less than)
- > (greater than)
- <= (less than or equal to)
- >= (greater than or equal to).

You can also use == for exactly equal to and != for not equal to. These comparisons can also be combined into more complex statements of logic using the words and and or.

We can add an else command to this code (03_06_if_else.py).

```
response = input("Enter a number: ")
number = int(response)
if number > 10:
    print("That's a big number")
else:
    print("That's a small number")
```

Notice that when we get to the else part of the code, we remove the indentation, back to the same level as the corresponding if. The print command to be run whenever the if condition is NOT true must be indented.

You can see the result of running this program in Figure 3-5.

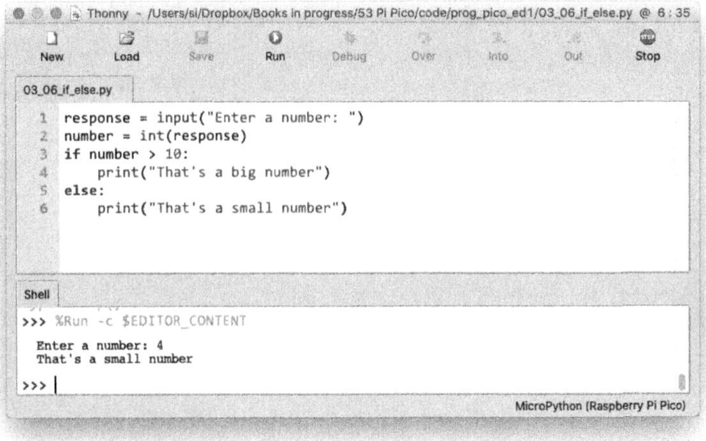

FIGURE 3-5 Running the if..else example in Thonny.

SUMMARY

This chapter has covered quite a lot of ground in terms of basic programming concepts. If you are new to programming it can take time to make sense of these ideas, so you may like to play with the code examples so far—alter them, run them, and see what happens. Experimenting is a great way to learn. If you mess up the programs, it doesn't matter you can just download them again.

In the next chapter you will learn all about one of the most important features of the Python language—functions.

4

Functions

Programs have a way of running away from you as they start to get more complex. Even with the simple SOS example in Chapter 3, you could see how the code was getting a bit repetitive and lengthy. Functions are one of the key tools that the programmer can use to make their code concise and easy to read.

You have already used some functions that are part of Python, such as `sleep`, `input` and `print`. However, you can and should also define your own functions when you are writing programs.

WHAT ARE FUNCTIONS?

A function is a little like a program within a program. When you make your own function, you give it a name and some code that should be run whenever you specify that the function should be run. Although programs and functions are often described as *running* when they are doing something, you will also hear the phrases *executing*, *calling* or *invoking* a function. They all mean the same thing.

A good program will have most of its code contained in functions, with just a few lines at the bottom (the eternal loop) where all the functions you defined earlier get used.

The code contained in a function can *call* or *invoke* other functions, and should be named in a way that reflects what it does. So function names usually describe some action, such as `blink` or `thow_dice`.

PARAMETERS

The built-in function `sleep` needs to know how long it needs to sleep for, and so takes a parameter in round brackets like

```
sleep(0.5)
```

which tells it to sleep for half a second.

Similarly, the function `print` expects a parameter of what should be printed

```
print("Pass the salt")
```

tells it to print the string inside the quotes.

Let's create our own version of `print` called `print_polite`. This function will simply add the string "please" to the end of any message printed. You can find this example in 04_01_print_polite.py.

```
def print_polite(text):
    print(text + " please")

print_polite("Pass the salt")
```

Run the program and you should see the text *Pass the salt please* in the Shell.

The word `def` (for define) marks the start of the function definition. An important point about functions is that just because they are defined, they are not necessarily used. Just because we have defined a function called `print_polite` does not mean that any of the lines of code within it will be run until we call it using the line

```
print_polite("Pass the salt")
```

After the word `def`, marking the start of the function definition, we have the function's name. This name must be unique within your program, and is how you will refer to the function when you come to call it. So, it can't be the same as any variables, or other words that Python uses like `print`, `if` etc. It follows the same naming convention as variables, so separate multiple words with an underscore character, as we have with `print_polite`. You also need to pick a name for your function that describes what it does, so that any reader of your program has a good idea what's going on.

Functions

Let's pickup our SOS example from the program 03_04_sos_for.py.

```python
from machine import Pin
from utime import sleep

led = Pin(25, Pin.OUT)

while True:
    # This loop does 3 dots (S)
    for x in range(1, 4):
        led.on()
        sleep(0.2)
        led.off()
        sleep(0.2)
    sleep(0.4) # Gap between S and O
    # This loop does 3 dashes (O)
    for x in range(1, 4):
        led.on()
        sleep(0.6)
        led.off()
        sleep(0.6)
```

We have already improved this over our original SOS example by using `for` loops to repeat some of the blinking. We could make this code a lot neater if there were a function called `blink` that had two parameters—the number of times to blink, and the delay between the LED changing from on to off and vice-versa. If we had such a function, then our eternal loop would just become

```python
while True:
    blink(3, 0.2)
    sleep(0.4)
    blink(3, 0.6)
```

Of course this won't work yet, because we need to define the function `blink`. This will look something like this

```python
def blink(times, delay):
    for x in range(1, times+1):
        led.on()
        sleep(delay)
        led.off()
        sleep(delay)
```

Notice how we have two parameters, `times` and `delay`, separated by a comma inside the parentheses. After this, there is a : to indicate the start of the function's code.

The first line of code in the `blink` function is a `for` loop that is going to control how many times the LED blinks. The `times` parameter is used here in the `range` command, with 1 added to it because the second parameter in range is exclusive.

Inside the `for` loop, are the four lines of code that turn the LED on and off with a `delay`. Here, the fixed number (0.2 or 0.6) that we used in the old code is replaced by the parameter `delay`.

Here is the full program (04_02_sos_function.py)

```
from machine import Pin
from utime import sleep

led = Pin(25, Pin.OUT)

def blink(times, delay):
    for x in range(1, times+1):
        led.on()
        sleep(delay)
        led.off()
        sleep(delay)

while True:
    blink(3, 0.2)
    sleep(0.4)
    blink(3, 0.6)
```

Notice that we have defined the `blink` function after the imports and the `led` variable definition. This is where you are expected to define any functions that the program uses—another convention that makes it easier for a newcomer to read your program.

RETURN VALUES

In the example above, the function does something—it makes an LED blink. When calling the function blink, we don't expect any value to come back from the function. In fact, Computer Scientists would say that blink is actually a *procedure* rather than a function for this very reason.

Python functions can *return* a value. That is, when you call them, you can assign the value they give back to a variable, or print it, or use it in some other way. To illustrate this, we could change the blink function so that when it's finished blinking, it returns the total duration that is spent doing blinking, calculated as the number of blinks, times the duration of each blink. Here's what it would look like

```
def blink(times, delay):
    for x in range(1, times+1):
        led.on()
        sleep(delay)
        led.off()
        sleep(delay)
    return times * delay * 2
```

The new line on the end of the function is highlighted. It starts with the return word, followed by a sum where the number of times blinked is multiplied by the delay and then again by 2 (there are two delays). We have now specified that our function will return a number, however, we are under no obligation to use that return value in our code. So, having added that line to the function, the program, 04_02_sos_function.py would still work just fine with this version of the blink function. However, it does mean that we can use the return value if we want to. For example, we could change the eternal loop to look like this

```
while True:
    print(blink(3, 0.2))
    sleep(0.4)
    print(blink(3, 0.6))
```

Now, the result of each call to blink will be printed in the Shell like this

```
1.2
3.6
1.2
3.6
1.2
3.6
1.2
```

You can find this modified program in 04_03_sos_return.py. This addition to the blink function isn't actually very useful, so lets look at another example of using return values, and introduce another useful Python module.

RANDOM NUMBERS

Computers behave in a very predictable manner. If you want to make them appear unpredictable, there is a Python library called random specifically for this purpose. Load the program 04_04_dice.py. Now that you have met a few Python programs, take a good look at it and try and predict what it will do when you come to run it.

```
from machine import Pin
from utime import sleep
from random import randint

led = Pin(25, Pin.OUT)

def throw_dice():
    return randint(1, 6)

def blink(times, delay):
    for x in range(1, times+1):
        led.on()
        sleep(delay)
        led.off()
        sleep(delay)

while True:
```

```
input()
dice_throw = throw_dice()
print(dice_throw)
blink(dice_throw, 0.2)
```

The areas of particular interest are highlighted. The first new thing is that we are importing the `randint` function from the `random` module.

We have copied the useful `blink` function from our earlier programs, and have also defined a new function called `throw_dice`. The `throw_dice` function just returns the result of calling the built-in function `randint` from the `random` module, with parameters of 1 and 6, which is the range of random numbers that we want. You may think that defining a function that only has one line of code is barely worth-while, and you'd have a point. However we will expand this function later and in fact, wrapping up the action of throwing the virtual dice in a function has potential advantages—if for some reason, we decided to use a twelve-sided dice, we would only have to change the code in the function rather than everywhere else.

The eternal loop starts with an `input` command. This has the effect that the program pauses until you press the Enter key in the Shell. Having pressed Enter, the function `throw_dice` is called and the result (a random number between 1 and 6) will be assigned to the variable `dice_throw`. This number is then printed and blinked.

Try pressing Enter a few times—your Pico should blink out the result of the dice throw, and the number should appear in the Shell. Wait until it finishes before hitting Enter again.

NAMED PARAMETERS

In the previous discussion, we said that we might want to change the function `throw_dice`. Let's change it now, so that we can simulate throwing more than one dice. It would be a nice touch if the number of dice to throw were an optional parameter so that, by default, just one dice would be thrown. Throwing two dice does not result in a total that is between 1 and 12 because the lowest throw you could have is a pair of 1s. We could calculate the throw as a random number between the number of dice and six times the number of dice, but let's make the code behave as if we actually throw three dice and add up the total. Here is the revised `throw_dice` function.

```
def throw_dice(num_dice=1):
    total = 0
    for x in range(1, num_dice+1):
        total += randint(1, 6)
    return total
```

The first thing to notice is the parameter to `throw_dice` now has a named parameter called num_dice. This has an =1 after it, which means that the default value for the parameter is 1. In other words, if you do not supply a parameter (when calling the function) the function will assume that you just want to throw one dice.

On a stylistic note, parameters like this are the only time I prefer to use = without a space on each side.

The first line of the function creates a new variable called `total`, which will be used to keep the total throw.

To throw two dice, we can change the code where we call throw_dice to look like this

```
 dice_throw = throw_dice(num_dice=2)
```

We could also just write

```
dice_throw = throw_dice(2)
```

but I think its clearer actually to name the parameter when calling the function.

You can find the full program for this in the file 04_05_dice_many.py.

SUMMARY

Now that we know how functions work, and how we can use them to make our programs more concise, we can start to develop some more complex programs.

In the next chapter, we will continue with the Morse code example, looking at how we can use lists and dictionaries ultimately to make ourselves a Morse code translator that, when given a sentence of text, will flash it out as dots and dashes.

5

Lists and Dictionaries

In this chapter, we will work towards creating a Morse code translation program, that will convert text that we type in the Shell into Morse code flashes of light. Along the way, we will learn how to use Python lists and dictionaries.

One way to think of programs is as virtual models of the real world, that you can then experiment with. So, if we want to create a Morse code translation program, we need to have a way of modelling Morse code, and then make the translation program work using this model.

To model the Morse code, we need to be able to look up the sequence of dots and dashes that form a given letter of a message. Python has a great way of representing such things in the form of *dictionaries*. However, before we move on to look at Python dictionaries, lets take a look at Python *lists*.

LISTS

So far, all of our variables have been single values: just one number or string of text. Most computer languages (including Python) have a way of representing a list of values. Sometimes they are called *arrays*, *vectors* or *collections*, but in Python they are just called *lists*. Figure 5-1 shows a pictorial representation of a list.

```
0 | 10
1 | 34
2 | 5
3 | 92
4 | 16
```

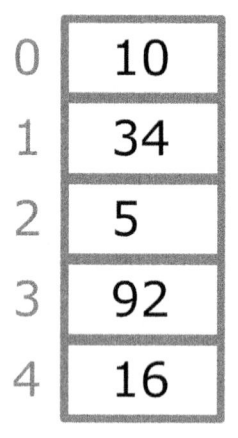

FIGURE 5-1 A list of numbers.

To create this list of numbers in Python, you would write the following code

`list_of_numbers = [10, 34, 5, 92, 16]`

Now, type this into the Python Shell and we can try some experiments on it.

```
>>> list_of_numbers = [10, 34, 5, 92, 16]
>>> list_of_numbers
[10, 34, 5, 92, 16]
>>>
```

When you type the variable name `list_of_numbers` for a second time, Python will display the numbers within square brackets.

To access individual elements from a Python list, you need to specify the position (also called *index*) of the element of the list that you want. Positions start at 0, so the first element of this list is at position 0 and you can retrieve it like this

```
>>> list_of_numbers[0]
10
>>>
```

If you wanted the second element of the list, you would type

```
>>> list_of_numbers[1]
34
>>>
```

Lists and Dictionaries

You can also change the value at a particular index position in the list like this

```
>>> list_of_numbers[0] = 1
>>> list_of_numbers
[1, 34, 5, 92, 16]
>>>
```

If you want to manipulate your list in other ways, such as sorting it, or reversing its order, you can do things like this

```
>>> list_of_numbers.sort()
>>> list_of_numbers
[1, 5, 16, 34, 92]
>>> list_of_numbers.reverse()
>>> list_of_numbers
[92, 34, 16, 5, 1]
>>>
```

You can also add new elements to the list as shown below.

```
>>> list_of_numbers = [10, 34, 5, 92, 16]
>>> list_of_numbers.append(12)
>>> list_of_numbers
[10, 34, 5, 92, 16, 12]
>>>
```

If you don't want the new element at the end of the list, then you can use insert which takes two parameters: the first being the position to insert the new element at and the second is the value to insert.

```
>>> list_of_numbers.insert(1, 12)
>>> list_of_numbers
[10, 12, 34, 5, 92, 16]
>>>
```

Note that insert(1, 12) will insert the number 12 as the second element—the numbering always starts from 0.

LIST EXAMPLE

There are lots more things you can do with lists but, for now, let's just say that lists are very flexible and move onto a more useful example. Later in this chapter, we will use a different type of data structure (dictionary) to make a general purpose Morse Code translator but, in the meantime, let's look at how we could use a list of time delays to blink out SOS (yet again). You can find the program in 05_01_sos_list.py.

```
from machine import Pin
from utime import sleep

led = Pin(25, Pin.OUT)

delays = [0.2, 0.2, 0.2, 0.6, 0.6, 0.6]

while True:
    for delay in delays:
        led.on()
        sleep(delay)
        led.off()
        sleep(delay)
```

Notice a difference in the `for` loop from how we used it previously. Now the `for` loop will give us each element in the list in turn rather than count by one each time.

STRINGS AS LISTS

You can treat a string as a list of individual characters. In fact, you can use the same square bracket notation to fetch individual characters from a string. Try the following in the Shell.

```
>>> s = "Raspberry Pi Pico"
>>> s[0]
'R'
>>>
```

Lists and Dictionaries

You can find the length of the string by doing

```
>>> len(s)
17
>>>
```

showing that spaces are treated just the same way as any other characters in the string.

As well as being able to access individual letters from the string, you can also select a range of letters in the string. For example, if we wanted the first three letters we could do

```
>>> s = "Raspberry Pi Pico"
>>> s[0:3]
'Ras'
```

The number before the : is the start index of the substring and the second number is one more than the end index.

If you put a negative number in after the :, instead of counting forwards, the code works from the end of the string, like this

```
>>> s = "Raspberry Pi Pico"
>>> s[0:-4]
'Raspberry Pi '
>>>
```

One difference between strings and lists is that you cannot change a string once it has been created. So, the example below will cause an error.

```
>>> s = "Raspberry Pi Pico"
>>> s[0] ='r'
Traceback (most recent call last):
  File "<stdin>", line 1, in <module>
TypeError: 'str' object doesn't support item assignment
>>>
```

DICTIONARIES

Whereas, in a list, you access things by their position in the list, or you take each element of the list in turn and do something with it, a dictionary is a little different. A dictionary holds a collection of values, but each value has a key associated with it that you use to access that *value*. This is rather like a telephone directory—the key is the person's name, and the value is their telephone number. For this reason, dictionaries are often called *lookup tables*.

They key is usually a string, but it doesn't have to be; it could be a number or any other type of data.

Let's define a dictionary with two entries in it. It's probably a good idea to follow along with these examples in the Python Shell.

```
>>> d = {'x' : 23, 'y' : 77}
>>> d
{'y': 77, 'x': 23}
>>>
```

The first thing to notice is that, whereas when you define a list you use square brackets, dictionaries are defined using curly-braces. We have defined a dictionary with two entries in it. Each entry is separated by a comma (just like lists) but, in a dictionary, each entry must have two parts separated by a : (the key and the value). So, the first entry has as key of the string 'x' with a value 23. The second entry in the dictionary has a key of 'y' and a value of 77.

Interestingly when we ask Python to display the value of our dictionary d, it does not show the two entries in the same order as when we created the dictionary. This is an important difference between dictionaries and lists. You cannot rely on the order of the entries in a dictionary staying the same. Figure 5-2 shows a graphical representation of this dictionary.

If we want to access a particular entry in the dictionary, we have a couple of choices. If we are certain that the key for the entry that we are going to ask for definitely exists, then we can use a square bracket notation like this

```
>>> d['x']
23
>>>
```

Lists and Dictionaries

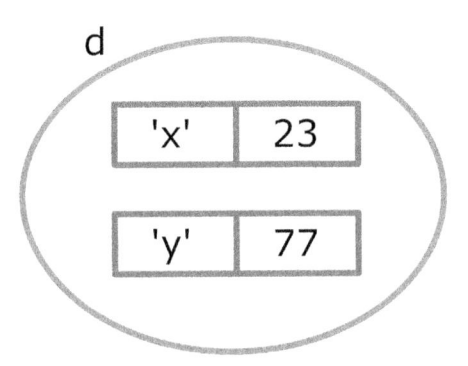

FIGURE 5-2 A Python dictionary.

However, if instead of asking for `'x'` or `'y'` (which exist) we asked for `'z'` (which doesn't) we will get an error like this

```
>>> d['z']
Traceback (most recent call last):
  File "<stdin>", line 1, in <module>
KeyError: z
>>>
```

So, for situations where the key might not be there, but we don't want to cause an error, we can use get like this

```
>>> d
{'y': 77, 'x': 23}
>>> d.get('x')
23
>>> d.get('z')
>>>
```

We haven't got an error, even though they key doesn't exists, but instead nothing came back. In fact, what came back was Python's special value of None. We can see this using print

```
>>> print(d.get('z'))
None
>>>
```

To add an new entry to the dictionary, or indeed to replace an entry with a particular key, you can just do this

```
>>> d['z'] = 1234
>>> d
{'y': 77, 'x': 23, 'z': 1234}
>>>
```

A MORSE CODE TRANSLATOR

Now that we know how to use dictionaries, we can create a dictionary to represent the Morse code – then, for any letter, we can lookup the sequence of dots and dashes to be flashed. You can find a table detailing the Morse code on its Wikipedia page (https://en.wikipedia.org/wiki/Morse_code).

Lets start by creating a dictionary for the letters A to Z.

```
codes = {
    "a" : ".-",    "b" : "-...",  "c" : "-.-.",
    "d" : "-..",   "e" : ".",     "f" : "..-.",
    "g" : "--.",   "h" : "....",  "i" : "..",
    "j" : ".---",  "k" : "-.-",   "l" : ".-..",
    "m" : "--",    "n" : "-.",    "o" : "---",
    "p" : ".--.",  "q" : "--.-",  "r" : ".-.",
    "s" : "...",   "t" : "-",     "u" : "..-",
    "v" : "...-",  "w" : ".--",   "x" : "-..-",
    "y" : "-.--",  "z" : "--.."
}
```

where the positioning of the curley braces and the layout are just for ease of reading.

You can find this dictionary in the file 05_02_morse_dict.py. The program won't do anything except assign the dictionary to the variable codes. So, run it and then try looking up various letters using the Shell like this

```
>>> codes['a']
'.-'
>>>
```

This is telling us that the code for the letter a is .- (dot dash).

Lists and Dictionaries

Let's now write a function called `send_morse_for` that takes a single letter and will eventually flash the Morse code for that letter. You can find this, in 05_03_morse_send_morse.py. This function doesn't do any flashing yet, it just uses `print` commands so that we can see what it does. Eventually, we will flesh out this skeleton of a function so that it flashes the LED. But, for now, we just want to print out the character chosen and its Morse representation.

```
def send_morse_for(character):
    if character == ' ':
        print("space")
    else:
        dots_n_dashes = codes.get(character.lower())
        if dots_n_dashes:
            print(character + " " + dots_n_dashes)
        else:
            print("unknown character: " + character)
```

Run the program and then enter the following in the Shell

```
>>> send_morse_for('z')
z --..
>>>
```

Let's take a look at what's going on. First we check to see if the character passed into the function for sending is a space character. If it is, we just print the string 'space' so we can see the program is working. This will later be replaced by a space-length pause.

If the character is not a space character, then the `codes` dictionary is used to lookup the `dots_n_dashes` to be flashed. We want the program to work for upper and lower case letters and so, when looking up the dots and dashes in the dictionary, the letter is converted to lowercase using the `lower` method.

If the character is in the dictionary (determined by `if dots_n_dashes`), the `print` command reminds us of the character and prints the sequence of dots and dashes. If, on the other hand, the character is not in the codes dictionary (it might be a number or punctuation) the missing character is printed.

Entering each character separately would definitely start to get a bit tedious after a while, and so let's expand the program so that it lets us type in a whole sentence. You will find this in 05_04_morse_printing.py. We have added an eternal loop containing code to prompt us to enter some text.

```
while True:
    text = input("Message: ")
    for character in text:
        send_morse_for(character)
```

Try running it and enter some text, as shown below. From what's being printed out things look very promising.

```
Message: Raspberry Pi Pico
R .-.
a .-
s ...
p .--.
b -...
e .
r .-.
r .-.
y -.--
space
P .--.
i ..
space
P .--.
i ..
c -.-.
o ---
Message:
```

Before we go any further, let's find out a bit more about the timings involved in Morse code.

A quick search on the web tells us that Morse code does not dictate the length of the dots in seconds, but it does state that the duration of a dash should be three times the duration of a dot. The gap between each letter of a message should also be three times the duration of a dot, and the gap between words should be seven dots-worth of duration. We can use this information to define some variables at the top of our program like this

Lists and Dictionaries

```
dot_duration = 0.2
dash_duration = dot_duration * 3
word_gap = dot_duration * 7
```

By defining the dash_duration and word_gap in terms of the dot_duration, should we want to speed up or slow down our sending of Morse code, all we need to do is change dot_duration, and the other time delay times will take care of themselves.

It's time now to move on to some flashing. We can make this happen by first adding the imports at the top of the file for Pin and sleep and define the LED like this

```
from machine import Pin
from utime import sleep

led = Pin(25, Pin.OUT)
```

Now, we need to modify the function send_morse_for so that, instead of just printing what its planning to flash, it actually does the flashing. Here is the modified function

```
def send_morse_for(character):
    if character == ' ':
        sleep(word_gap)
    else:
        dots_n_dashes = codes.get(character.lower())
        if dots_n_dashes:
            print(character + " " + dots_n_dashes)
            for pulse in dots_n_dashes:
                send_pulse(pulse)
            sleep(dash_duration)
        else:
            print("unknown character: " + character)
```

The first thing that happens is that, if the character being flashed is a space (signifying the end of a word) there is no flashing to do—but we do need to delay for word_gap seconds. On the other hand, if the character is not a space, then we look-up the dot and dash pattern for the character in the codes dictionary. If the character is not there (maybe the message included numbers or other characters not in the codes dictionary) then we print a message to that effect, but if it is there, then we need to flash the pattern of dots and dashes. So, a for loop is used to step over each dot or dash in turn and then call a function (that we haven't written yet) called send_pulse. This new function receives a single '.' or '-' as its parameter, and does a short or long flash depending on whether its a dot or a dash.

```
def send_pulse(dot_or_dash):
    if dot_or_dash == '.':
        delay = dot_duration
    else:
        delay = dash_duration
    led.on()
    sleep(delay)
    led.off()
    sleep(delay)
```

The send_pulse function sets the value of the variable delay, depending on whether the character is a dot or a dash, and then does a single flash using that delay time.

You can find the full Morse program in 05_05_morse_complete.py. Run it, and type some messages into the Shell (Figure 5-3). You should now see the message being flashed out on the Pico's built-in LED.

```
from machine import Pin
from utime import sleep

led = Pin(25, Pin.OUT)

dot_duration = 0.2
dash_duration = dot_duration * 3
word_gap = dot_duration * 7

durations = {"." : dot_duration, "-" : dash_duration}

codes = {
    "a" : ".-",     "b" : "-...",   "c" : "-.-.",
    "d" : "-..",    "e" : ".",      "f" : "..-.",
    "g" : "--.",    "h" : "....",   "i" : "..",
```

Lists and Dictionaries

```
        "j" : ".---",  "k" : "-.-",   "l" : ".-..",
        "m" : "--",    "n" : "-.",    "o" : "---",
        "p" : ".--.",  "q" : "--.-",  "r" : ".-.",
        "s" : "...",   "t" : "-",     "u" : "..-",
        "v" : "...-",  "w" : ".--",   "x" : "-..-",
        "y" : "-.--",  "z" : "--.."
}

def send_pulse(dot_or_dash):
    if dot_or_dash == '.':
        delay = dot_duration
    else:
        delay = dash_duration
    led.on()
    sleep(delay)
    led.off()
    sleep(delay)

def send_morse_for(character):
    if character == ' ':
        sleep(word_gap)
    else:
        dots_n_dashes = codes.get(character.lower())
        if dots_n_dashes:
            print(character + " " + dots_n_dashes)
            for pulse in dots_n_dashes:
                send_pulse(pulse)
            sleep(dash_duration)
        else:
            print("unknown character: " + character)

while True:
    text = input("Message: ")
    for character in text:
        send_morse_for(character)
```

As an exercise, you might like to try adding in the extra number characters specified in the Morse code (see https://en.wikipedia.org/wiki/Morse_code). You might also want to try speeding things up by changing the value of dot_duration.

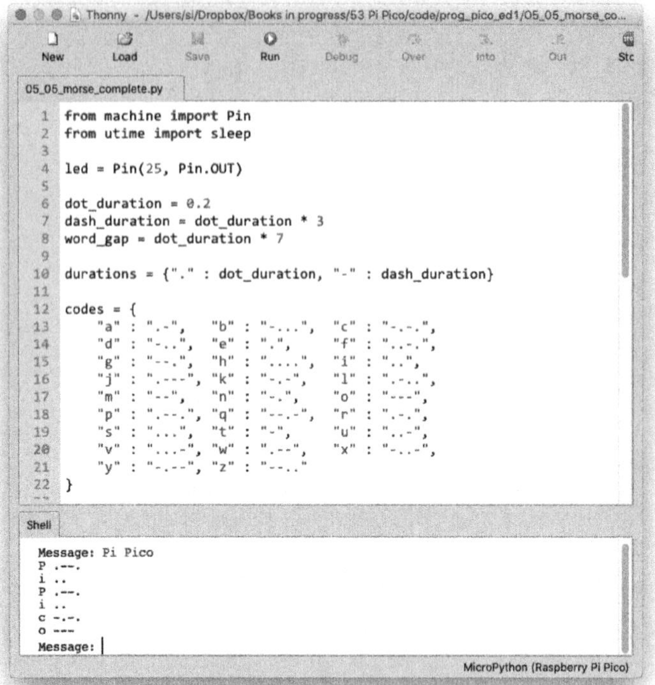

FIGURE 5-3 Morse code translator.

SUMMARY

In this chapter we have learnt how useful lists and dictionaries can be. We could have made our Morse program using a huge series of `if` statements, checking for each letter in turn and then flashing the dots and dashes, but this would have resulted in a program hundreds of lines long. By using a dictionary, we have kept our program to a nice concise 50 or so lines of code.

In the next chapter, we will look at some more ideas to keep our code neat and tidy.

6

Modules, Classes and Files

We have been importing modules into our programs since almost the start of this book, so, it is finally time to take a look at exactly what's going on here and introduce the concept of classes and object-oriented programming.

This chapter is getting a little advanced, so don't worry if not all of it makes senseon first reading, you may find yourself coming back to it as you progress through the book.

IMPORTING FROM MODULES

MicroPython includes lots of ready-made functions for us to use in our programs. If every ready-made function were immediately available to us, then there would be a real risk that (when thinking up the name of our own functions) a function with that name already exists, if there are two functions with the same name, how will Python know which one to use when you call the function?

Try running the following program

```
def print():
    pass

print()
```

In Python, pass does not actually do anything, but is required in situations like this, where there would otherwise be an empty block of code.

When you run this, you will see a somewhat cryptic error message like this

```
ManagementError
>>> THONNY FAILED TO EXECUTE COMMAND get_globals


SCRIPT:
__thonny_helper.print_mgmt_value({name : (__thonny_helper.repr(value),
id(value)) for (name, value) in globals().items() if not
name.startswith('__')})

STDOUT:


STDERR:
Traceback (most recent call last):
  File "<stdin>", line 1, in <module>
  File "<stdin>", line 21, in print_mgmt_value
TypeError: function takes 0 positional arguments but 3 were given
```

This is Python's way of telling us that we can't give our own function the name print, because that name is already in use.

The function print is available to use in programs without having to do anything special to include it. Other functions (the vast majority) are hidden away in modules, and we can only use them in our programs by importing them like this

```
from utime import sleep
```

This tells Python that we want to use the sleep function from the utime library. Once we have made the import at the top of the program, we can simply call sleep anywhere in your programs like this

```
sleep(0.5)
```

If we were feeling adventurous, we could just import everything from the utime library in one go by writing

```
import utime
```

The down side to importing everything is that, unless we have an encyclopedic knowledge of Python, we don't know exactly what functions will be imported.

DOCUMENTATION

We know `sleep` lives in `utime`, but we have no idea what other functions we might be importing at the same time with `import utime`, and if we only need `sleep`, then it's safest just to import `sleep`. We can actually see what else is available in `utime` either by consulting the documentation for `utime` (https://docs.micropython.org/en/latest/library/utime.html) or by running this command.

```
>>> import utime
>>> dir(utime)
['__class__', '__name__', 'gmtime', 'localtime', 'mktime', 'sleep', 'sleep_ms', 'sleep_us', 'ticks_add', 'ticks_cpu', 'ticks_diff', 'ticks_ms', 'ticks_us', 'time', 'time_ns']
>>>
```

It's possible to see what some of these functions do from their names, but to make proper use of them we may need to consult the documentation. The documentation for MicroPython is available at https://docs.micropython.org/ Be aware, though, that this is not specific to the Raspberry Pi Pico version, and not all modules will be available, and equally importantly, the Pico specific machine library is documented separately at https://datasheets.raspberrypi.org/pico/raspberry-pi-pico-python-sdk.pdf.

USEFUL BUILT-IN MODULES

MicroPython has a large number of useful modules that can save you time with your programs. There is no sense in reinventing the wheel, especially when someone who is really good at making wheels will give you one for free!

You can find a list of all the modules at the MicroPython documentation here: https://docs.micropython.org/en/latest/library/index.html.

Note that often the words *library* and *module* are used interchangeably. Also, because this Python is running on a Pico, memory space is at a premium. If you are used to using full-strength Python, you will find that not everything in every module is implemented. However, all the most useful stuff should be there.

Here are some highlights of the most used built-in modules.

random

In our dice example, earlier in the book, we met the `randint` function from the `random` module. We used this function to generate a random number in a range, like this

```
>>> randint(1,6)
4
>>>
```

Another useful function in the random module is `choice`. You can use this function to randomly select one of the entries in a list like this

```
>>> from random import choice
>>> fruit = ['apple', 'banana', 'orange', 'grape']
>>> choice(fruit)
'apple'
>>> choice(fruit)
'grape'
>>>
```

While you could just pick a random number between 0 and the length of the list minus 1, and then access the element of the list at that index position, using `choice` is nicer.

math

You get all the basic arithmetic operators like +, -, * and / (and some others) without having to use a module. However, if you need more advanced functions from trigonometry, logarithms etc., you have to use the math module.

The mathematical constants e and ∏ are available.

```
>>> from math import pi
>>> pi
3.141593
```

However, it is important to know that all the trigonometric functions in the `math` module use `radians` rather than degrees. There is a handy function radians that will do the conversion. So, for instance, to find the sin of 30 degrees you could do this

```
>>> from math import radians, sin
>>> sin(radians(30))
0.5
>>>
```

Another math function that is quite useful is `pow`. This raises one number (int or float) to the power of another. For example

```
>>> from math import pow
>>> pow(2, 16)
65536.0
>>>
```

There are many more functions in this module and, if you find that there is something mathematical that you need to do, check first—there will probably already be a function for it.

CLASSES

Python is (like most recent languages) *object-oriented*. There are varying degrees of object-orientedness in languages and, in the case of Python, it's not radically object-oriented. Let's look at what this means.

If you recall, back in Chapter 5, we wanted to prepare messages to be sent by Morse code by converting them to lowercase. We did this using `lower()` like this

```
>>> "AbC".lower()
'abc'
>>>
```

A lowercase string with the same letters has been created for us from the mixed case string 'AbC'. However, you may be asking yourself why the dot, and why that order. Why aren't we treating `lower` as a function and writing the following (which won't work by the way)

```
>>> lower("AbC")
```

especially as we can write

```
>>> len("AbC")
3
>>>
```

This really highlights Python's fairly lackadaisical approach to object-orientation. Sometime you use the dot notation, sometimes you use a function call. So, lets take a look at what's going on with the object-oriented way of doing things.

```
>>> "AbC".lower()
'abc'
>>>
```

The formal way we would describe the use of the dot here is to say that we are calling the method `lower` on the *instance* of the class string "AbC". Let's unpack that and define a few things.

First of all strings. There might be lots of strings in use in a program, and they all have the thing in common—they consist of a list of characters. You could say that they are all of the same type or class. In fact, we can ask Python to tell us the class of something like this (note that's double underscore before and after the word class)

```
>>> "abc".__class__
<class 'str'>
>>> [1,2,3].__class__
<class 'list'>
>>>
```

"abc" is said to be an *instance* of the class `str`—and the thing about a class is that, as well as saying something about the type of data it contains, it can also

Modules, Classes and Files

contain functions that are particular to that class (such as `lower` in this example). Functions like `lower` that are linked to a class are called *methods*. We can see all the methods available on the class `str` like this

```
>>> dir(str)
['__class__', '__name__', 'count', 'endswith', 'find', 'format',
'index', 'isalpha', 'isdigit', 'islower', 'isspace', 'isupper',
'join', 'lower', 'lstrip', 'replace', 'rfind', 'rindex', 'rsplit',
'rstrip', 'split', 'startswith', 'strip', 'upper', '__bases__',
'__dict__', 'encode']
>>>
```

As you can see, there's `lower` and the corresponding `upper` and a load of other methods—some of which are more useful than others.

Modules often contain a class with lots of methods. We have already met this in the form of the `machine` library, from which we imported `Pin` in order to be able to make an LED blink. Here's a reminder, where we have simplified program 03_01_blink just to turn the LED on. If you don't want to type this in, you can find it in the file 06_02_Pin_class.py.

```
from machine import Pin
led = Pin(25, Pin.OUT)
led.on()
```

You may wonder why `Pin` has a capital P. That's because `Pin` is the name of a class and class names, by convention, start with an uppercase letter. So, when we import `Pin`, we are not importing a single function like we did when we imported `randint` from `random`, but, we are importing an entire class called `Pin`.

We can find out what `Pin` is capable of (what's in it) by doing this

```
>>> dir(Pin)
['__class__', '__name__', 'value', '__bases__', '__dict__', 'ALT',
'IN', 'IRQ_FALLING', 'IRQ_RISING', 'OPEN_DRAIN', 'OUT', 'PULL_DOWN',
'PULL_UP', 'high', 'init', 'irq', 'low', 'off', 'on', 'toggle']
>>>
```

The methods `on` and `off` are there, and a lot of other things that we don't need to know about, as well as some others that we will meet in later chapters.

You can define your own classes and methods, but you rarely need to do this when writing small programs for the Pico, so we will consider this an advanced topic and leave it for other resources to cover. You can find a good explanation of classes in Python at https://realpython.com/python3-object-oriented-programming/.

STRING METHODS

In uncovering how our use of the `lower` method worked, we have discovered other methods that we can use on strings, some of which are very useful. While we are here, lets go through some of these.

find

This method returns the index position for the starting point of one string within another. If the string being sought isn't in the string being searched, then this method returns –1. For example

```
>>> "Raspberry Pi Pico".find("Pi")
10
>>>
```

In this example we are searching for the string *Pi* within *Raspberry Pi Pico*. This method is often used to find out if one string contains another or not, by seeing if it returns –1.

format

Whether displaying messages in the Shell, or perhaps on a display attached to a Pico, a common problem is the need to display a combination of a message and some values. The `format` method provides a handy way of doing this. It uses a template string that contains markers indicated by curly braces. The parameters supplied in the format method are then substituted into the template string in order. For example

```
>>> "The temperature is {} degrees C".format(23.4)
```

```
'The temperature is 23.4 degrees C'
>>>
```

'You can substitute more than one parameter. Each must be sparated by a comma and you must have the same number of {}markers as parameters.

You can also achieve the same result by concatenating strings using +. However, each thing that you add must be a string, and often numbers will need to be converted into strings using the `str` function.

```
>>> "The temperature is " + str(23.4) + " degrees C"
'The temperature is 23.4 degrees C'
```

Which approach you take is entirely a matter of personal taste.

replace

You can create a new string from an existing string, with some of the text replaced using the `replace` method. For example, if you wanted to change all occurrences of b in a string to B you could do it like this

```
>>> "abcb".replace("b", "B")
'aBcB'
```

split

If you have a string that contains a file path such as: /home/pi/myfiles/file1.txt, and you want each part of the file path as its own entry in a list, then the `split` command will save you a lot of effort.

In the example, the file path is first split on the / character to find its parts. The last part (the actual file name) is then split on . to separate the base file name from the extension.

```
>>> parts = "/home/pi/myfiles/file1.txt".split("/")
>>> parts
>>> parts[:-1]
['', 'home', 'pi', 'myfiles']
>>> file = parts[-1]
>>> file
'file1.txt'
```

```
>>> file.split(".")
['file1', 'txt']
```

Notice how `[-1]` is used as a shorthand to access the last element of the list.

strip

A common problem when dealing with a string that a user has typed in – perhaps from an input command – is that there are unwanted spaces on either end of what they typed. The strip method removes those spaces without touching any spaces between words in the string. Here's an example that assumes the user has typed in their name is a rather sloppy manner.

```
>>> "   Simon Monk   ".strip()
'Simon Monk'
```

FILES AND EXCEPTIONS

The Raspberry Pi Pico, has some 2MB of flash storage, that has to be shared between the program running on the Pico and any data that you might want to store on the Pico. 2MB might not sound like very much, but for a small microcontroller board, this is well above average. This means that you have some space available for storing data on the Pico itself. This can be very useful in projects like data-loggers, where you store sensor readings to a file periodically, and then when the logging is finished, copy the file of readings onto your computer.

Being able to save data persistently to a file, and then read it from the program, is also very useful when you want your program to remember things after a reset. For example, if you had a program that operated in different modes (perhaps turning something on and off) you might want to remember what state it was in when the Pico was powered down, when you power the Pico back up again. Another example would be to remember a high-score in a game.

We will start by creating a file called test.txt, and writing the number 1 into it. You can find this program in 06_03_file_write.py.

Modules, Classes and Files

```
f = open("test.txt", "w")
f.write(str(1))
f.close()
```

The file is opened in write mode (`"w"`) and, when you are finished writing to the file, you should `close` it (note the use of the dot notation for the `close` method). We can only write strings into this file, and so the number 1 has to be converted into a string using `str`.

Run the program and then, in the Thonny *View* menu, enable *Files*. You should now see two lists on the left of the Thonny window for files on your computer and on the Pico. You should see the file test.txt there (see Figure 6-1). Click on it, to open it, and you should see that it does indeed contain the number 1. It is important to remember that the file is actually on your tiny little Pico, not on your computer. You can edit the files, so change 1 to 2 and then, click the Save icon.

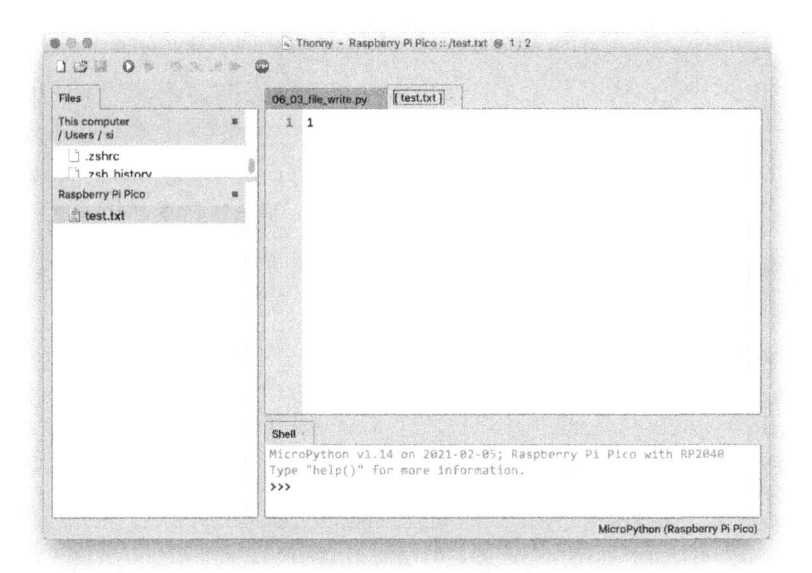

FIGURE 6-1 The files view in Thonny.

Now that we have a file on the Pico to use, let's write a program to read it. You can find this in 06_04_file_read.py.

```
f = open("test.txt", "r")
print(f.read())
f.close()
```

This time, the files is opened in read (`"r"`) mode.

You should see output in the Shell like this

```
>>> %Run -c $EDITOR_CONTENT
2
```

EXCEPTIONS

Reading a file like this is fine if we know that the file already exists, but a common scenario in reading a file on the Pico is that, the very first time the program is run, an attempt is made to read a non-existent file. If the file doesn't exist, then we want to create it rather than have an error occur.

The way to do this in MicroPython is to try and read the file and, if we get an error, then we know the file doesn't exist. So far, when we have had errors (such as when accessing a dictionary with an unknown key) the program has stopped (crashed) because of the error. In MicroPython we can catch such errors when they occur, and handle them in our code without it causing the program to crash. To do this, we use `try except`, as the example in 06_05_try_except.py.

```
file_name = "testXX.txt"
try:
    f = open(file_name, "r")
    print(f.read())
    f.close()
except:
    print("File {} doesn't exist".format(file_name))
```

Notice that the filename has been changed to testXX.txt and, because there is no such file on the Pico, we will see the message to that effect. The point is that the program did not crash, we are catching any error using the `try except` statement and handling the error ourselves.

FILE COUNTER EXAMPLE

Putting the read and write file programs together, the program 06_06_counter.py will keep track of how many times the Pico has been reset and its program run.

```
config_file = "config.txt"
count = 0

def read_config():
    global count
    try:
        f = open(config_file, "r")
        count = int(f.read())
        f.close()
    except:
        pass

def write_config():
    f = open(config_file, "w")
    f.write(str(count))
    f.close()

read_config()
print(count)
count += 1
write_config()
```

The program starts by defining two variables. The name of the file is held in config_file—this idea of a configuration file is common in embedded programming. The other variable (count) will be used to hold the count of times the program has been run.

Inside the read_config function the line global count indicates to Python that the variable count is defined globally, outside of the function. You can read global variables from within a function (as we do in the write_config function) but if you want to change their value, you need a line like this at the top of the function.

The `print` statement below `except` has been replaced with `pass`. That is because Python insists on there being something inside `except`—but if you don't want to do anything, you indicate this by using `pass`.

After the two function definitions, we have the code that will be run just once before the program finishes. This reads the config file, prints the value of `count` that it reads, adds one to `count`, and then saves it again.

Run the program a few times and you will see something like this in the Shell.

```
>>> %Run -c $EDITOR_CONTENT
0
>>> %Run -c $EDITOR_CONTENT
1
>>> %Run -c $EDITOR_CONTENT
2
>>> %Run -c $EDITOR_CONTENT
3
>>>
```

If you now look in the file section of Thonny (after clicking *Refresh* in the little drop-down menu in the file area) you should see a new file config.txt and, inside it, you should see the current count.

SUMMARY

In this chapter we have learnt a bit more about modules, and how to use them, as well as the sometimes confusing style that results from Python's implementation of object-oriented programming. We have also explored the use of the Pico's spare flash memory for use as a file system.

In the next chapter, we will start using the Pico's GPIO pins.

7

Inputs and Outputs

In this chapter we will look at programming the Pico's GPIO pins. Rather than wade into the depths of electronics, we will restrict ourselves to nothing more complex than a metal paper-clip (or short length of wire) and the Pico's built-in LED. This way, we can be ready to embrace some proper electronics when we come to Chapter 8.

DIGITAL OUTPUTS

We have already used digital outputs quite extensively in controlling the built-in LED. This LED is connected to pin GP25 of the Pico's processor chip. This pin is not brought out to one of the connections running down the sides of the board, but it works in the same way. That is, the voltage of the pin can be set to be either 0V or 3.3V. If you have something (say an LED) connected to that pin, then you can turn it on by setting the pin to 3.3V or off by setting it to 0V. It is actually a bit more complicated than this, as we will discover in Chapter 8 when we come to connect an external LED to a pin, but this will do for now.

The reason the pins are called *GP* pins is that they are *general purpose*—that is, they can act as inputs or outputs. Before we can use a pin as a *digital output*, we need to tell the Pico that this is how we intend to use it, like this

```
from machine import Pin
led = Pin(25, Pin.OUT)
```

As we discussed in Chapter 6, `Pin` is a class and, if you type `Pin.OUT` into the shell, it will tell you that it has the value 1. So `Pin.OUT` makes it clear that the pin is to be used as an output than having to remember that 1 means output (and by the way `Pin.IN` is 0). Now that we have a reference to the pin called `led`, and we have told the Pico to configure it as an output, we can turn it on and off using various methods on the `Pin` class.

The methods `on` and `high` both set the output to 3.3V (lighting the LED) and the corresponding `off` and `low` set it to 0V (turning it off). It is entirely a matter of personal choice whether you use `on`/`off` or `high`/`low`.

```
>>> led.high()
>>> led.low()
```

As well as the methods that explicitly set the output high or low, there is the `toggle` method—this sets the output high if its low, and to low if its currently high. Try running this command a few times in the Shell. We could use this to make our LED code simpler.

```
>>> led.toggle()
>>> led.toggle()
>>> led.toggle()
```

TIP: In the Shell, pressing your up arrow cursor key will recall the last command you entered in the Shell, so that you don't have to type the whole line again.

There is a third way to set a digital output to be high or low, and that is using the `value` method. This method takes a single parameter—0 for low (0V) or 1 for high (3.3V).

```
>>> led.value(0)
>>> led.value(1)
```

DIGITAL INPUTS

Whereas for a digital output, the Pico is controlling something, turning it on or off, a digital *input* is used to detect whether a switch of some sort is on or off. A

Inputs and Outputs

digital input is *read*, to see whether the voltage at the pin is above or below a threshold roughly half-way between 0V and 3.3V.

You should not be surprised to hear that this is how we tell a Pico to use a particular GP pin (10 in this case) as an input

```
>>> switch = Pin(10, Pin.IN)
```

A convention, when connecting switches to a digital input, is to use what is called a pull-up resistor (Figure 7-1) connected to the digital input, that biases the input towards 3.3V (high). The switch is then connected between the digital input and GND (0V), so that when the switch is pressed the digital input is connected to GND. Think of the pull-up resistor as a spring that pulls the GPIO pin high unless pulled the other way more strongly by the switch.

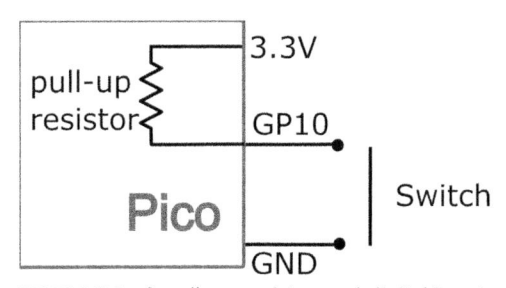

FIGURE 7-1 A pull-up resistor and digital input.

This means that when the switch is not pressed the digital input is at 3.3V (high) and when the switch is pressed, the input becomes 0V (low). This can be confusing as, logically, *high* usually means *on* rather than off—but, that's fine, we can just compensate for this in the code.

We don't actually have to use a physical pull-up resistor when connecting a switch, because the Pico's GP pins all have one built-in that can be enabled and disabled from your code. So, when connecting a switch to a Pico you write the following—both to set the pin as an input and enable the built-in pull-up resistor.

```
switch = Pin(10, Pin.IN, Pin.PULL_UP)
```

The optional third parameter `Pin.PULL_UP` switches in the pull-up resistor.

We can test out digital inputs by running the program below (07_01_digital_input.py).

```
from machine import Pin
from utime import sleep

switch = Pin(10, Pin.IN, Pin.PULL_UP)

while True:
    print(switch.value())
    sleep(0.1)
```

When you see a stream of 1s appear in the Shell. This is because the input is being pulled-up high—here 1 means high (3.3V) or at least half-way to 3.3V and definitely not 0V.

Notice the use of `switch.value()` to determine whether the input is high (1) or low (0).

Now, as shown in Figure 7-2, use an unfolded paper-clip or other metallic item to connect GP10 to the convenient GND pin just above it. You should see that you now get 0s appearing in the Shell.

FIGURE 7-2 A make-shift switch using a paperclip.

Inputs and Outputs

Let's now look at another example of using a digital input. The program 07_02_switch_led.py, will light the built-in LED for 10 seconds when our paperclip switch connects GP10 to GND.

```
from machine import Pin
from utime import sleep

switch = Pin(10, Pin.IN, Pin.PULL_UP)
led = Pin(25, Pin.OUT)

while True:
    if switch.value() == 0:
        led.on()
        sleep(10)
        led.off()
```

Try running the program. Initially, the built-in LED will be off, but when you momentarily connect GP10 and GND using the paperclip, the LED will light and stay lit for 10 seconds.

A paperclip is not an ideal way of making a switch and, in the next chapter, we will explore the use of proper push-switches connected to digital inputs.

ANALOG OUTPUTS

Digital outputs allow us to turn things on and off— for example, we can light an LED or turn it off—but we can't use a digital output to change the brightness of an LED in a graduated way. That is where analog outputs come into play.

Analog outputs on the Pico should, more accurately, be described as *Pulse Width Modulation* (PWM) outputs. They are actually digital outputs that control the length of pulses that they generate. This can be used to control the brightness of an LED or the speed of a motor. Figure 7-3 shows how PWM works.

Thinking, for example, of an LED. If the repeating pulses are quite short, then the LED is only lit for a short amount of time and will appear to be dim. If the pulses are longer, then the LED will appear brighter. Since the pulses are so fast (by default 1000 pulses per second) your eye can't keep up, and you can't see the LED flickering.

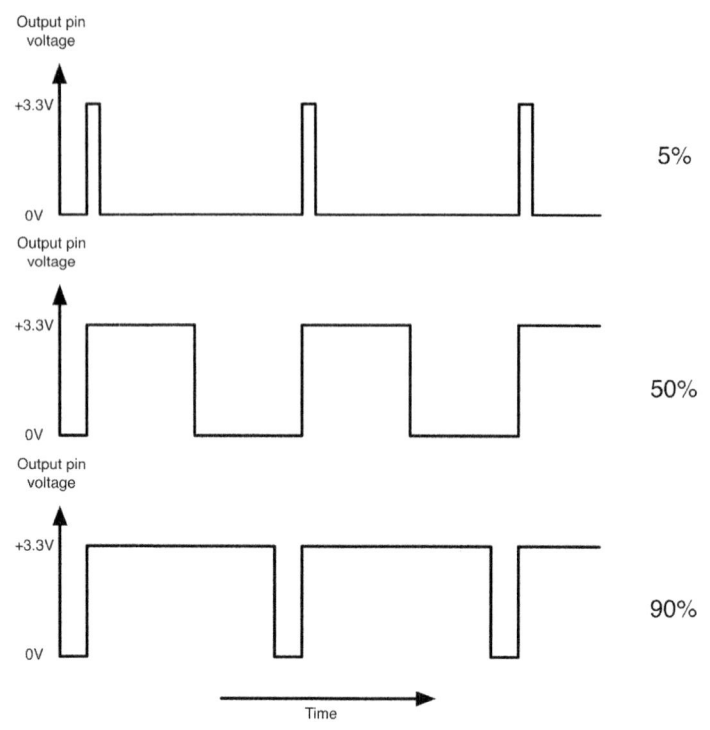

FIGURE 7-3 Pulse Width Modulation

You can use PWM on any of the Pico's pins, so let's try controlling the brightness of the built-in LED. Try running the program 07_03_pwm.py.

```
from machine import Pin, PWM

led = PWM(Pin(25))

while True:
    brightness_str = input("brightness (0-65534):")
    brightness = int(brightness_str)
    led.duty_u16(brightness)
```

This program prompts you to enter a number in the Shell between 0 and 65534. Try entering some numbers and notice how the brightness of the built-in LED varies.

```
brightness (0-65534):0
brightness (0-65534):65534
brightness (0-65534):32000
brightness (0-65534):5000
brightness (0-65534):2000
brightness (0-65534):1000
brightness (0-65534):0
brightness (0-65534):
```

When specifying how long the pulse is high (the *duty*) we enter a number between 0 and 65534. This number is 2 raised to the power of 16, minus 2.* When specifying that pin 25 is to be used for PWM output, we wrap the Pin definition in a PWM class. Remember that the input function returns a string, so this has to be converted to the number expected by duty_u16 using the int function. The function duty_u16 is so called because it expects a 16 bit unsigned number.

Over 65,000 steps of brightness is excessively precise for an application like this, so we could modify the program to that of 07_04_pwm_percent.py, where we enter a number between 0 and 100% brightness.

```
from machine import Pin, PWM
led = PWM(Pin(25))

while True:
    brightness_str = input("brightness (0-100):")
    brightness = int(int(brightness_str) * 65534 / 100)
    led.duty_u16(brightness)
```

Note that we have had to use int twice. Once to convert the text entered into a number and then again to convert the floating point number resulting from the multiplication and division into the integer value expected by duty_u16.

Generating pulses like this can also be used to control servomotors. We will see how to do this in the next chapter.

---

* It is not explained in the Pico's documentation why this is 2 rather than the 1 that you would expect.

ANALOG INPUTS

Digital inputs from something like a switch are either on or off – but, what if we want to measure the voltage at the pin, rather than simply detect if it is over the threshold voltage or not. Such a voltage might come from a sensor that we are using to measure temperature or light.

Figure 7-4 shows the difference between digital and analog inputs. As you can see, when you read a digital input, the value you get back is just 0 or 1 depending on whether the voltage at the pin is above or below the threshold voltage. Whereas with an analog input the reading should range between 0 and 65535. The way the Pico's analog inputs work, mean that the lowest reading you can get from an analog input is actually offset to 336, which is why the Analog Value line does not exactly intersect the chart's origin.

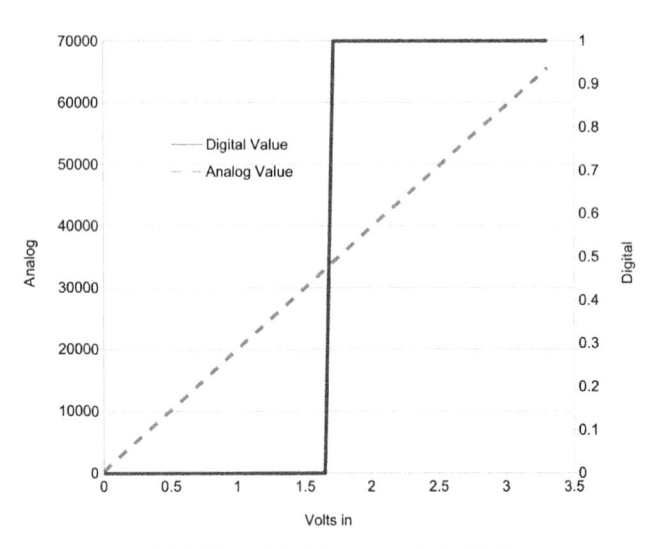

FIGURE 7-4 Digital and analog inputs

Try running program 07_05_analog_input.py and you should see a stream of numbers appear in the Shell. There will be some fluctuation as the analog input will be picking up electrical noise.

Inputs and Outputs

```python
from machine import ADC, Pin
from utime import sleep

analog = ADC(28)

while True:
    reading = analog.read_u16()
    print(reading)
    sleep(0.5)
```

Just like we wrapped `Pin` in `PWM` to use PWM, when we use an analog input the pin must be wrapped in the `ADC` (Analog Digital Converter) class. Note that, unlike digital inputs and outputs, analog inputs are only available on certain pins -- namely, GP26, 27 and 28. There is also an internal analog pin 4 that is connected to an on-board temperature sensor.

Let's modify the program a little so that it displays the actual voltage at the pin rather than the raw reading. You will find this in program 07_06_voltmeter.py.

```python
from machine import ADC, Pin
from utime import sleep

analog = ADC(28)

def volts_from_reading(reading):
    min_reading = 336
    max_reading = 65534
    reading_span = max_reading - min_reading
    volts_per_reading = 3.3 / reading_span
    volts = (reading - min_reading) * volts_per_reading
    return volts

while True:
    reading = analog.read_u16()
    print(volts_from_reading(reading))
    sleep(0.5)
```

When you run this program, the voltage at GP 28 will be printed in the Shell every half second. To convert the reading into a voltage, we first find the *span* of possible readings by subtracting the minimum reading from the maximum possible reading. The number of reading values per volt is then calculated as

Programming the Pico

3.3 divided by the span of readings. Finally, we can calculate the volts by first subtracting the minimum reading and then multiplying the result by `volts_per_reading`. The function is not very efficient, as all the calculations leading up to the calculation for `volts_per_reading` could actually just be made once. So, if performance is a problem, then move these lines outside the function so that they are run just one time.

We can check that this crude voltmeter is working by carefully using out paper clip to connect GP28 to 3.3V as shown in Figure 7-5.

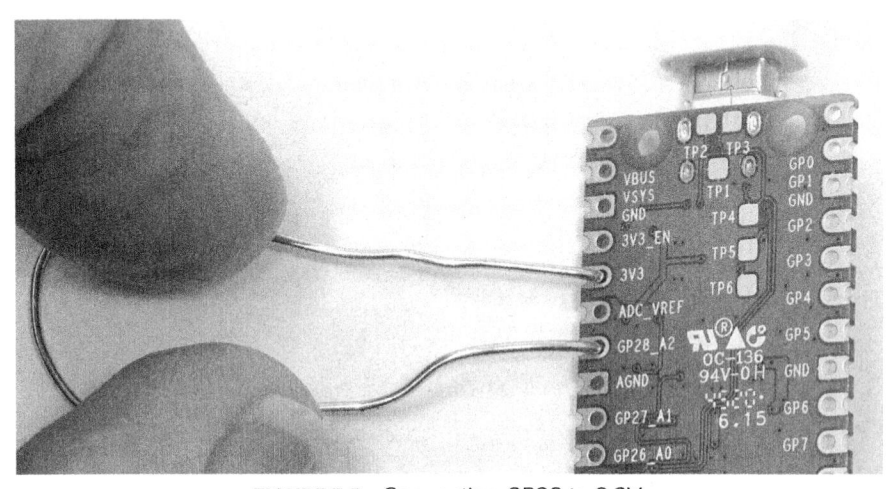

FIGURE 7-5 Connecting GP28 to 3.3V.

When you do this, you should see values of 3.3V appearing in the Shell.

To check the other end of the range, connect GP28 to GND as shown in Figure 7-6 and the readings should go to 0.

Inputs and Outputs

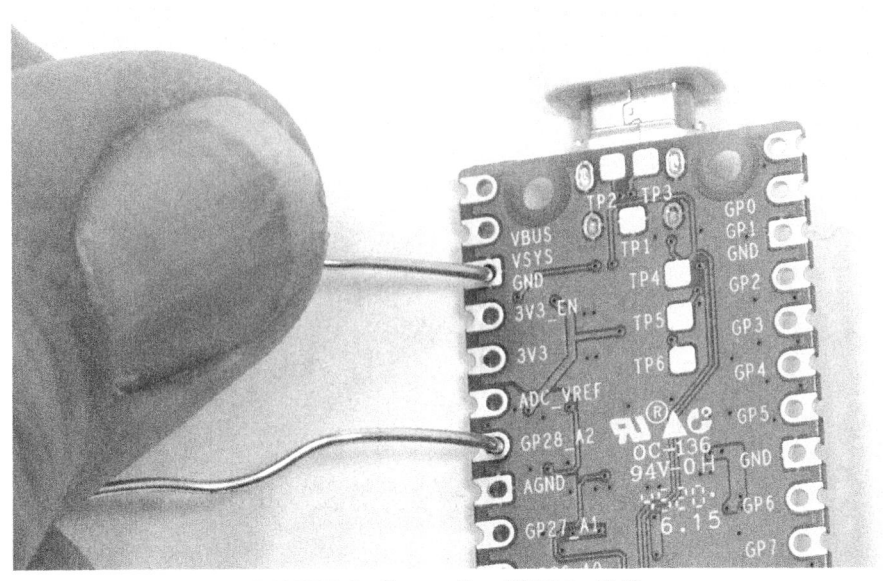

FIGURE 7-6 Connecting GP28 to GND.

SUMMARY

This chapter has dealt with the basics of inputs and outputs, but the Pico's input/output features are actually very advanced for such a low cost microcontroller board so, in Chapter 9, we will return to the subject of input and output and look at some of its more advanced features.

In this chapter we have taken using the built-in LED and a paperclip switch as far as we can to illustrate the basics of inputs and outputs. In the next chapter, we will start using the Pico with some external electronic components, and learn how to make simple electronic add-ons using solderless breadboard.

8

Electronics

This chapter builds on what you have learnt about analog and digital inputs and outputs, and uses this knowledge to construct some simple projects using a solderless breadboard (Figure 8-1) to connect electronic components to your Pico.

FIGURE 8-1 An example project using solderless breadboard.

In this chapter, there is no getting away from the need to have some actual electronic components available. MonkMakes Ltd manufactures a kit (the MonkMakes Electronics Kit 1 for Raspberry Pi Pico—see https://monkmakes.com/pico_bb) tailored for this chapter. This will give you everything you need to get started with electronics and the Pico.

SOLDERLESS BREADBOARD

Solderless breadboard is a great way to try out electronic circuits. The components (including the Pico) are pushed into metal clips behind the plastic face of the breadboard, so there is no need for any soldering—and it is easy to make different projects reusing the same components in different configurations.

As you can see, in Figure 8-1, a Raspberry Pi Pico is plugged into solderless breadboard, along with some other components that we will discuss later. *Jumper wires* are used to connect different parts of the circuit together.

Behind the face of solderless breadboard (let's just call it breadboard from now on) you will find metal clips (Figure 8-2) that connect together electrically, any component legs that are pushed through from the front.

The main area of the breadboard is divided into two columns of clips that connect all the holes of a particular row together. At the left and right side of the breadboard are long vertical clips called power buses. These can be used for any connections that you want to make, but are particularly useful for GND and 3.3V power connections that often need to be connected to lots of places on the breadboard—and much less likely that you will make a connection to the wrong pin.

Each row of the breadboard is normally marked with a number (0 to 30) and each column with a letter (a to j), helping to identify positions on the breadboard. The special MonkMakes breadboard included in the MonkMakes Electronics Kit 1 for Pico that accompanies this book (Figure 8-3) changes the row names for the top 20 rows to the names of the Pico pins, making it much easier to see how the Pico is connected.

Electronics

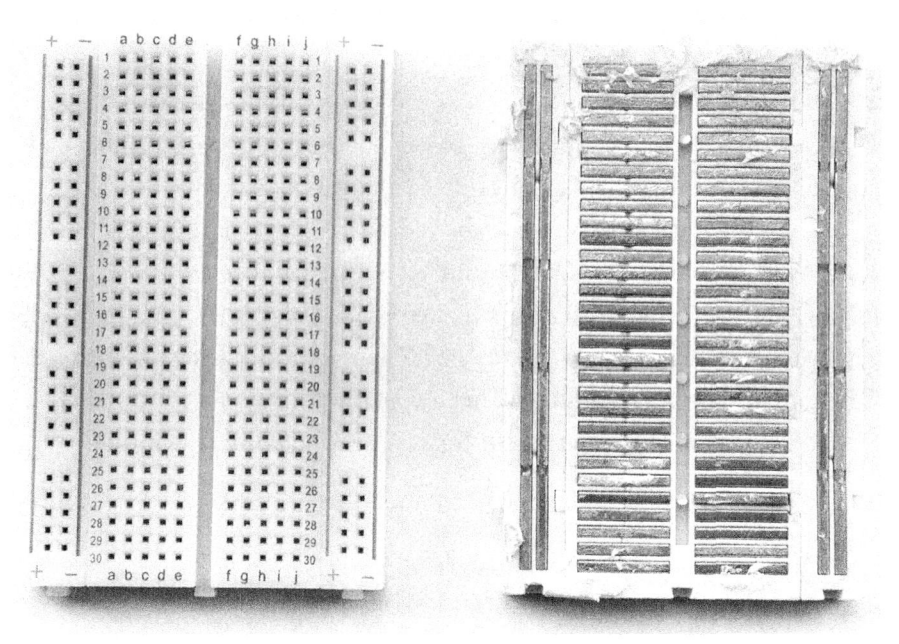

FIGURE 8-2 Breadboard deconstructed.

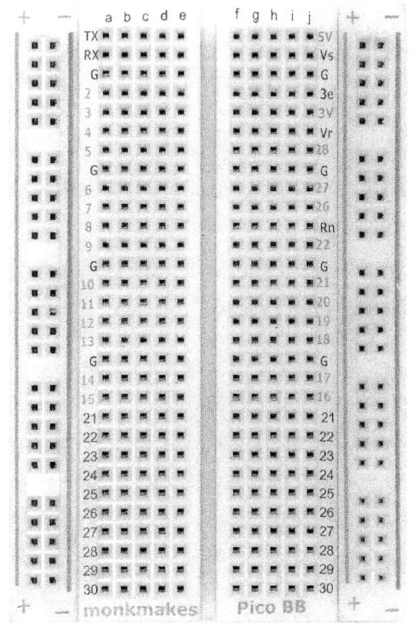

FIGURE 8-3 The MonkMakes Breadboard for Pico.

SOLDERING PINS ONTO YOUR PICO

To use the Pico with the breadboard, it needs to have header pins attached. Your Pico may already have had pins soldered onto it when you bought it. If not, you will need to solder header pins (included in the MonkMakes kit).

To keep the pins straight, it is a good idea to push their long ends into the solderless breadboard. This has the added advantage that once soldered, the Pico will be in the right place for making projects, and won't need to be relocated. It's a good idea to use a ruler or other flat object to push the rows of header pins onto the breadboard, as they can be sharp on your fingers (Figure 8-4).

The pins need to both start at the top row and run down columns c and h.

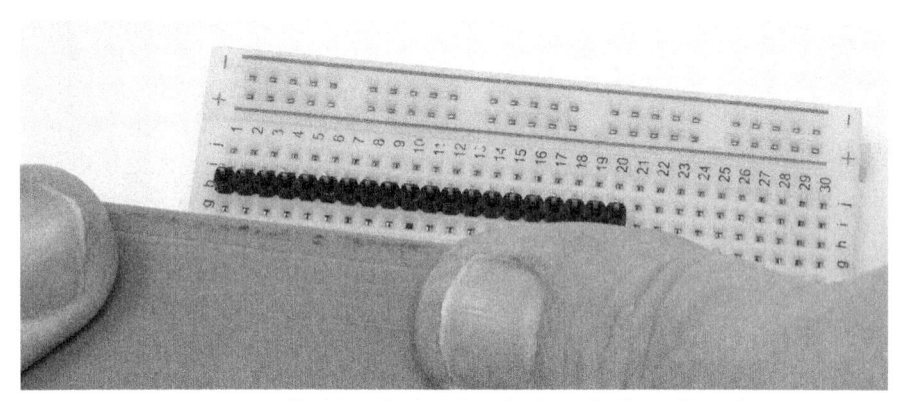

FIGURE 8-4 Pushing the header pins into the breadboard.

Once the pins are in the right place on the breadboard, the Pico should fit easily over the pins with the USB connector towards the edge of the breadboard (Figure 8-5).

If this is your first attempt at soldering, watch a few tutorials before you start to get an idea of what you are trying to do. Here's a good one: https://www.youtube.com/watch?v=37mW1i_oEpA.

Solder the pins, heating the pin where it meats the circuit board and then applying solder until it flows (Figure 8-6).

Electronics

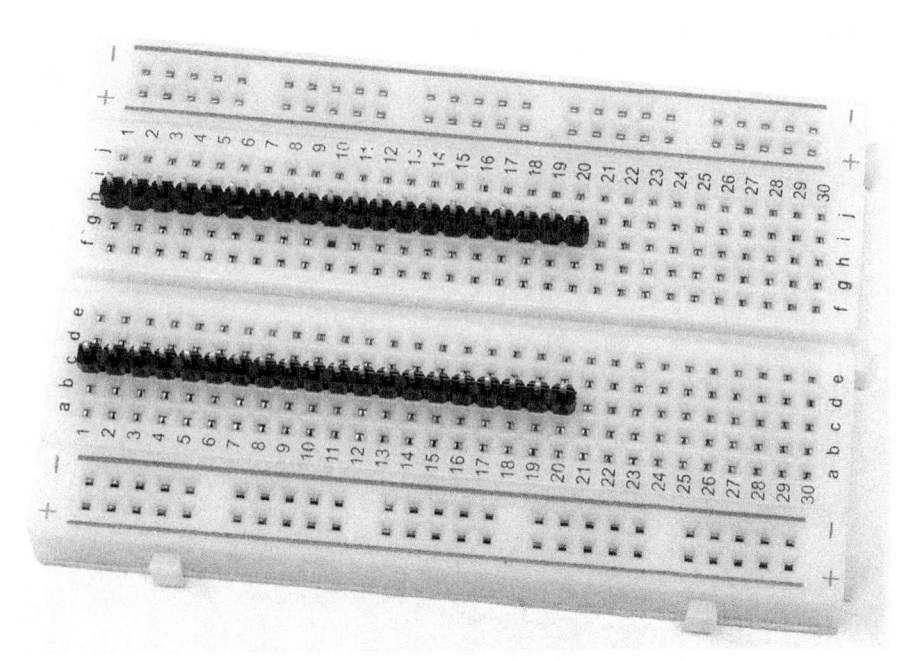

FIGURE 8-5 The header pins in columns c and h.

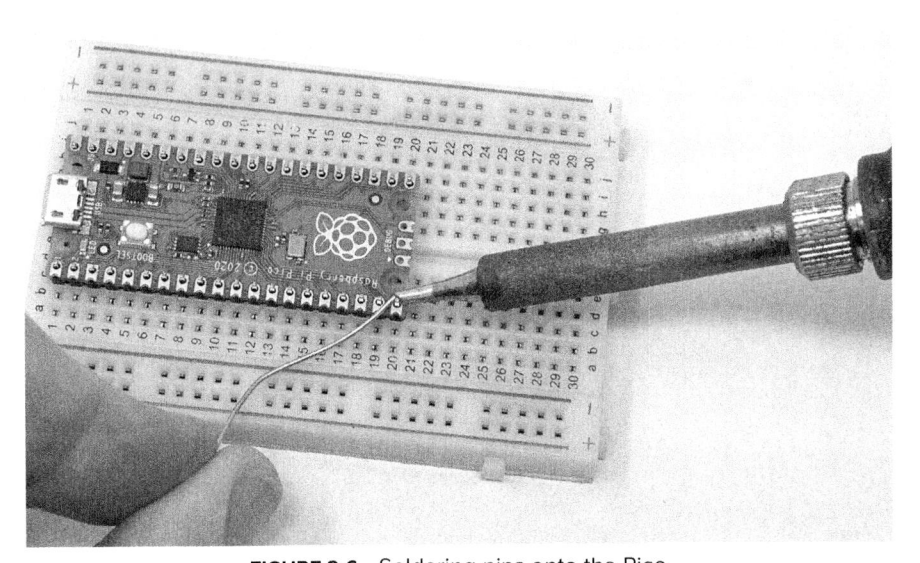

FIGURE 8-6 Soldering pins onto the Pico.

Once all the pins are soldered, your breadboard should look like Figure 8-7.

FIGURE 8-7 The Pico on breadboard.

COMPONENTS

The MonkMakes Electronics Kit 1 for Raspberry Pi Pico includes a useful set of components to get you started with your electronics projects. The kit also includes its own booklet describing a number of projects that you can make with the kit. Table 8-1 introduces these components.

Electronics

TABLE 8-1 Parts Included in the MonkMakes Electronics Kit 1 for Pico

Jumper wires—use these to connect things together on the breadboard.	
Servomotor and bag of servomotor arms and screws.	
470Ω resistors (yellow, violet, brown stripes). Resistors for use with LEDs, to limit the current.	
1kΩ resistors (brown, black, red). For use with the phototransistor to measure light intensity.	
4.7MΩ resistor (yellow, violet and green stripes). To make a touch sensor.	
Red LEDs—the longer lead is the + (positive) lead.	
RGB LED—the longest lead is the common negative lead. A red, green, blue multi-color LED.	
Tactile push switches.	
Phototransistor. This is used for measuring light intensity. The longer lead is the negative lead.	
Buzzer. This makes a noise when connected to an oscillating pin on the Pico.	
Variable resistor (pot). You can read the position of the knob using an analog input on the Pico.	

We will meet most of these components as we progress through the book.

MAKING LEDS BLINK

Having made the Pico's built-in LED blink in the last chapter, lets now attach an external red LED to the Pico and make that blink instead. Figure 8-8 show a schematic diagram for this.

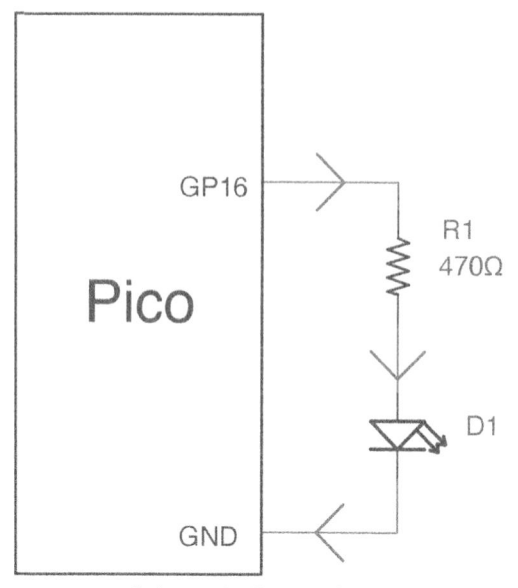

FIGURE 8-8 Schematic diagram for a Pico and LED.

We are going to connect the LED to pin GP16. When GP16 is high (3.3V) current will flow through the resistor (R1—zig-zag line) through the LED, and back into the Pico's GND connection—making a circuit that causes the LED (D1) to light as the current flows through it. The resistor is needed to reduce the current flowing through the LED—too much current though the LED shortens the LEDs life, and could also damage the GPIO pin that the Pico is connected to.

Connect the components to the breadboard as shown in Figure 8-9. Note that the LED has a positive and negative lead, the positive lead is a bit longer than the negative lead and should go to row 27 (connecting it to the resistor). The 470Ω resistor has yellow, violet and brown stripes and it does not matter which way around this goes.

Electronics

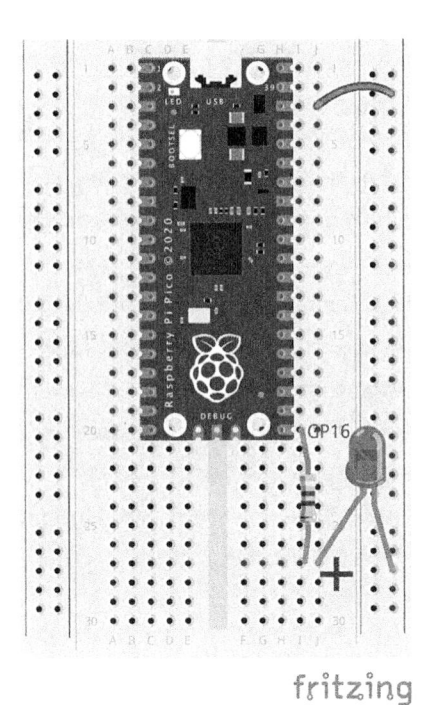

FIGURE 8-9 The wiring diagram for an LED.

Load the program 08_01_blink_led.py into Thonny and run it. The LED should start to blink.

```
from machine import Pin
from utime import sleep

led = Pin(16, Pin.OUT)

while True:
    led.on()
    sleep(0.5) # pause
    led.off()
    sleep(0.5)
```

The code is the same as we used to blink the internal LED, however, we are now specifying pin 16 (external LED) instead of 25 (built-in LED).

RGB LEDS

RGB (red, green, blue) LEDs are made up of three LED colors in one LED body. By varying the amount of red, green and blue (using PWM) it is possible to change the color of the LED. We can write a simple program that push switch to select from a list of different colors to cycle through.

Figure 8-10 show the schematic diagram of our circuit, where R1, R2, R3 are three resistors, and S1 is a push switch. Given that LEDs have two legs each you might expect the package to have six legs—but actually it just has 4. This is because the negative ends of the red, green and blue LEDs are connected to the same pin on the package (the longest leg).

Each LED channel must have its own resistor to limit the current and the push switch connects Pico GP12 to GND (0V) when the button is pressed.

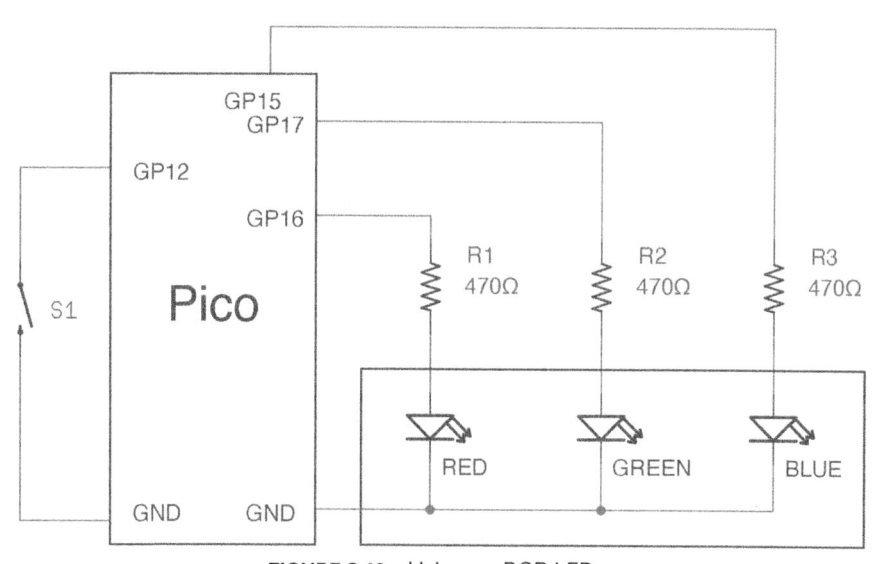

FIGURE 8-10 Using an RGB LED.

Start by connecting up the components as shown in Figure 8-11. All three of the resistors used in this project are the same value of 470Ω. You should find that one of the RGB LED's leads is longer than the rest.

Place the RGB LED such that the long lead is in row 28 on column j of the breadboard and connected to the GND (black) wire. Make sure that the legs of the resistors are not touching each other.

The push switch should be inserted with its legs on rows 28 and 30 on column e.

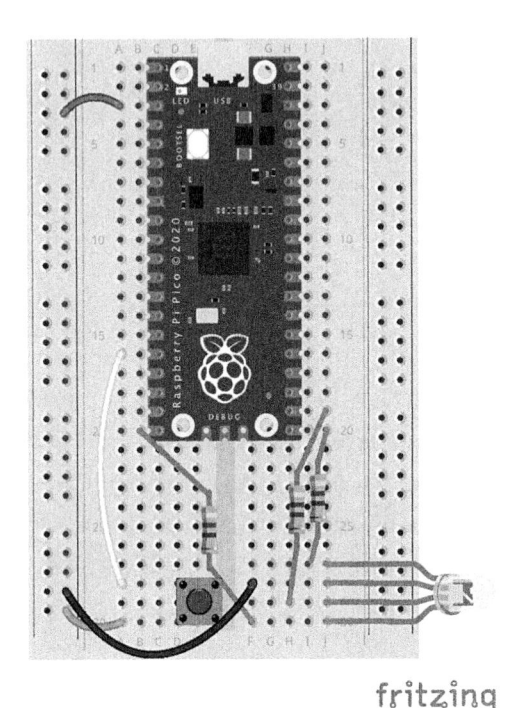

FIGURE 8-11 The wiring diagram for an RGB LED and switch.

The program for the RBG LED can be found in the file 08_02_RGB.py.

```
from machine import Pin, PWM
from utime import sleep

red_ch = PWM(Pin(16))
green_ch = PWM(Pin(17))
blue_ch = PWM(Pin(15))

button = Pin(12, Pin.IN, Pin.PULL_UP)
colors = [[255, 0, 0], [127, 127, 0],
```

```
                [0, 255, 0], [0, 127, 127],
                [0, 0, 255], [127, 0, 127]]
def set_color(rgb):
    red_ch.duty_u16(rgb[0] * 256)
    green_ch.duty_u16(rgb[1] * 256)
    blue_ch.duty_u16(rgb[2] * 256)

index = 0
set_color(colors[index])
while True:
    if button.value() == 0:
        index +=1
        if index >= len(colors):
            index = 0
        sleep(0.2)
        set_color(colors[index])
```

This code makes use of a list of lists (`colors`) to represent the different colors that the LED displays. Each of the inner lists is a list of three values, in order—the red, green and blue brightness, as a number between 0 (off) and 255 (fully on). For example [127, 0, 127] means red half brightness, green off and blue half brightness—making a kind of purple color.

The `set_color` function takes such an array and sets the three PWM duty-cycle values, at the same time, multiplying the value by 256, so that it spans the full range of PWM brightnesses.

The eternal loop checks the button and, if it is pressed (`button.value() == 0`) then 1 is added to `index`. Once `index` becomes equal to or greater than the number of colors in the list `colors`, then we need to wrap around to the start of the list by setting `index` to 0. Finally, after handling the button press, we delay for 0.2 of a second; this delay prevents switch bouncing, where pressing the switch once may register as several quick presses, as the switch makes repeated contacts as the button is being pressed. This stops some color values appearing to be skipped when the button is pressed. It also means that, if you hold the switch, the LED will cycle through the colors.

SERVOMOTORS

Servomotors (Figure 8-12) differ from most motors, in that their intended use is to set the position of an arm attached to their rotor to a particular angle (roughly in the range 0 to 180 degrees) rather than to rotate continuously.

FIGURE 8-12 A small servomotor.

A servomotor is actually quite a sophisticated little device—Figure 8-13 shows a representation of what's inside one.

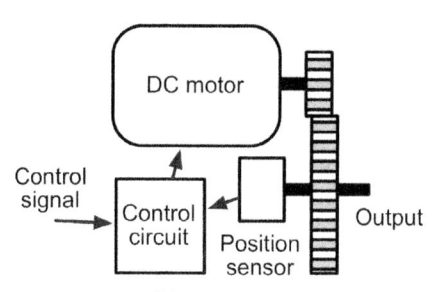

FIGURE 8-13 How a servomotor works.

A DC motor (a normal motor, that rotates continuously) is attached to a gear box, which drives the servo arm is also coupled to a position sensor that provides feedback to an electronic control circuit so that the servo arm is kept at the correct angle.

Control of the position uses a series of pulses arriving every 20 milliseconds. Figure 8-14 shows how different pulse lengths result in the servo arm moving to a different position.

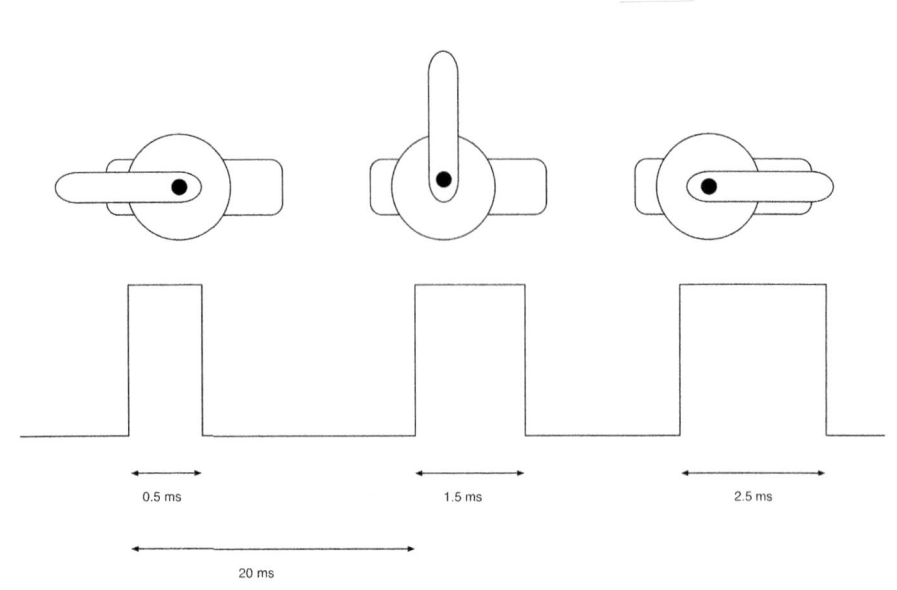

FIGURE 8-14 Servomotor control pulses.

A short pulse, of just 0.5 milliseconds, will put the arm at one end of its travel. A pulse of 1.5 milliseconds will put the arm at its center position, and a pulse length of 2.5 milliseconds will put it at the other end of its travel.

Before we look at how we are going to write code to generate these pulses, lets wire up a circuit to experiment with. Figure 8-15 shows the schematic for our test circuit—we will use two switches—one to decrease the angle of the servomotor's arm and another to increase it.

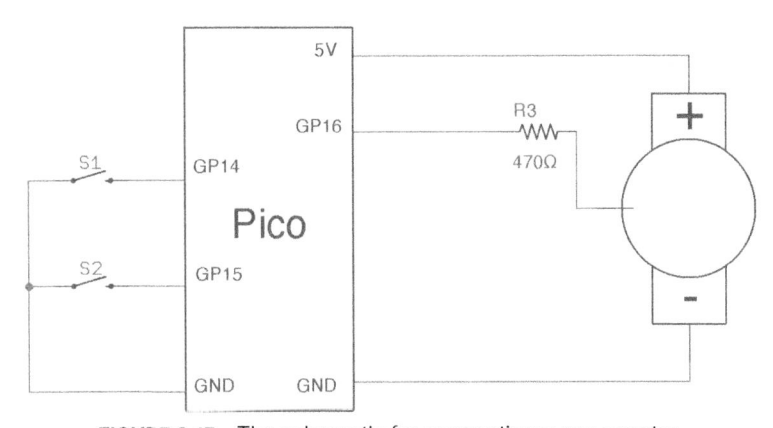

FIGURE 8-15 The schematic for connecting a servomotor.

Electronics

Although the servomotor included in the kit will work at 3V, more power is available if it is powered from the 5V available over the USB lead – that is the power connection that we use here. The control connection to the servo is connected via a 470Ω resistor. While not strictly essential this does mean that, if the servomotor tries to draw too much current from the GPIO pin, the Pico will not be damaged. Figure 8-16 shows the breadboard layout.

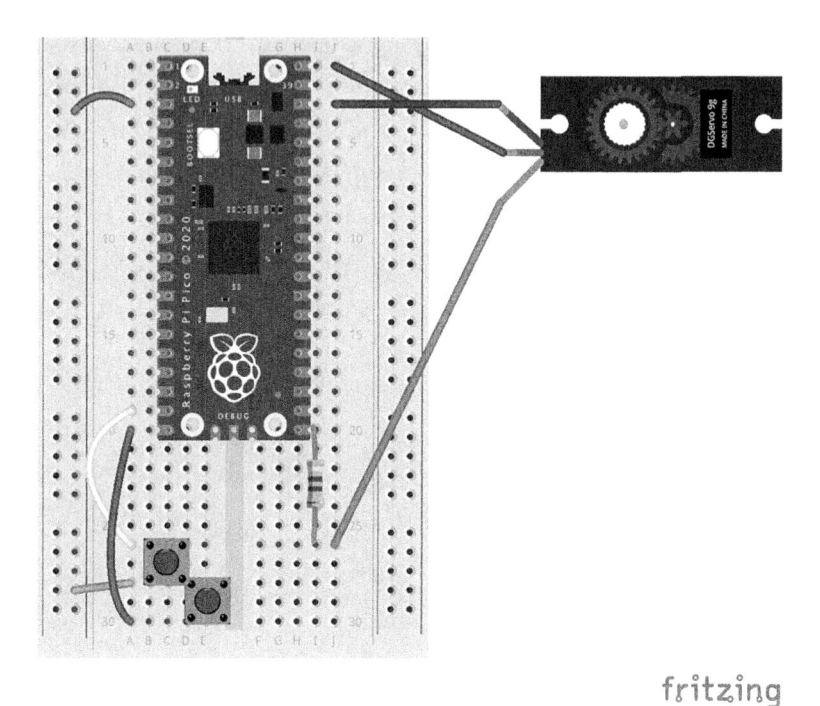

FIGURE 8-16 The wiring diagram for a servomotor and switches.

To connect the servomotor, use jumper wires. The red lead of the servomotor should be connected to 5V on the Pico, the brown lead of the servomotor to GND, and the orange lead to row 26—as shown in Figure 8-16. Insert the switches carefully—note that they share one row of connectors on the breadboard.

The servomotor is supplied with a little bag of arms that fit over the cogged drive shaft of the servo motor. Select the one shown in Figure 8-12, and push it onto the drive shaft. It doesn't matter at what angle you attach it, as you can always adjust it later.

Open the file 08_04_servo.py in Thonny, and run it. You should find that, when you press one button, the servo arm rotates one way; when you press the other button, it should rotate the other way. You will also see the angle printed out in the Shell.

If the servo arm is not free to move throughout its range you can adjust the position of the arm by taking it off and repositioning it. To get it at the straight up 90 degree position, stop the program and then run it again. This will start the servo arm at the middle 90 degree position. Then, without pressing either button, carefully take the arm off and reposition it so that it is pointing straight up—as shown in Figure 8-12. Note that, because of the cogged drive shaft, you probably won't be able to get the arm at exactly the 90 degree position—anything close is fine.

Here's the code

```
from machine import Pin, PWM
from utime import sleep

button_up = Pin(14, Pin.IN, Pin.PULL_UP)
button_down = Pin(15, Pin.IN, Pin.PULL_UP)

servo = PWM(Pin(16))
servo.freq(50) # pulse every 20ms

def set_angle(angle, min_pulse_us=500, max_pulse_us=2500):
    us_per_degree = (max_pulse_us - min_pulse_us) / 180
    pulse_us = us_per_degree * angle + min_pulse_us
    # duty 0 to 1023. At 50Hz, each duty_point is
    # 20000/65535 = 0.305 µs/duty_point
    duty = int(pulse_us / 0.305)
    print(angle)
    servo.duty_u16(duty)

angle = 90
set_angle(90)
min_angle = 10
```

```
max_angle = 160

while True:
    if button_up.value() == 0 and angle <= max_angle:
        angle += 1
        set_angle(angle)
        #print(angle)
        sleep(0.01)
    elif button_down.value() == 0 and angle > min_angle:
        angle -= 1
        set_angle(angle)
        #print(angle)
        sleep(0.01)
```

The PWM frequency is set to 50Hz (pulses per second) (`servo.freq(50)`), as that is the frequency of pulses that the servomotor expects.

Most of the work in generating the pulses to control the servomotor is contained in the function `set_angle`. This means that, if you want to create your own projects using servomotors, you can just copy this function.

The most important parameter for `set_angle` is the angle that you want to set the servomotor's arm to, which should be between 0 and 180. The other parameters set the minimum and maximum pulse width—you should not need to change these unless you have an unusual servomotor. The function starts by working out the number of microseconds of pulse which will be required for each degree of angle. It then calculates the total pulse length in microseconds for the angle required. Finally, it calculates the duty (the PWM value) between 0 and 65535, and uses `servo.duty_u16` to set that pulse width on the servomotor's control pin.

The eternal loop waits for button presses and changes the angle accordingly. Since many servomotors can't manage the full 180 degrees, the range of angle set is limited to between 10 and 170. If you find that the servomotor judders badly at one end of its travel, you may need to adjust `min_angle` or `max_angle`.

SUMMARY

In this chapter, we have explored the use of solderless breadboard to make some basic electronic circuits. In the next chapter, we will look at some more advanced features of the Pico's GPIO pins.

9

Advanced Inputs and Outputs

Now that you have done a little experimenting, attaching some electronics to your Pico, let's explore some more advanced features of the Pico that you might find useful.

The Pico's IO (Input Output) capabilities far outstrip most microcontrollers. These include high speed PWM, IO pins that can have their own simple processing hardware and, perhaps most impressively, there are two cores (processors) in the Pico's RP2040 microcontroller.

INTERRUPTS

So far, when we have needed to detect that a button has been pressed, we have put some code like this example from 08_02_RGB.py in the eternal loop

```
button = Pin(12, Pin.IN, Pin.PULL_UP)

while True:
    if button.value() == 0:
        # do something
        sleep(0.2) # debouncing
```

This is called *polling*, because we just keep checking the state of the digital input over and over again until we read a 0, indicating a button press. This is fine for a simple program like this, where nothing else is happening, but imagine the situation where we were doing something time-consuming in the loop—perhaps reading a sensor value and displaying it on a display. In that case there would be a risk that a very quick button press might happen when the code was busy doing something else, and be missed.

For situations like this, you can use *interrupts*. A digital input can be specified as causing an interrupt. As the name suggests, when an interrupt is detected whatever code is currently being run is suspended and an interrupt service routine (ISR) is run. When the ISR has finished, the program continues where it left off. Lets have a look at an example of this.

This interrupt is triggered by the change in input voltage of a digital input— for example when a switch is pressed.

To try this out, connect a switch between GP12 and GND, using the breadboard layout of Figure 9-1.

FIGURE 9-1 A switch to test interrupts.

Advanced Inputs and Outputs

Now run the program 09_01_interrupts.py.

```
from machine import Pin
from utime import sleep

button = Pin(12, Pin.IN, Pin.PULL_UP)

def handle_button(ignore):
    print("BUTTON PRESSED")

button.irq(handle_button, Pin.IRQ_FALLING)

i = 0

while True:
    i += 1
    print(i)
    sleep(0.2)
```

The code that we want to be run following an interrupt is placed in a function. The `irq` (interrupt request) method on `button` associates the interrupt handler function (`handle_button` in this case) with the `button` pin.

We want the interrupt to happen when the input pin button goes from high to low (i.e. when it is first pressed) and so the second parameter to `irq` is set to `Pin.IRQ_FALLING`. If you wanted the interrupt to happen when the button was released, rather than when it was pressed, then you would specify `IRQ_RISING`. Try changing it and notice the effect of the change when you run the program.

If your project requires more than one interrupt, then that is fine, you can associate multiple pins with an interrupt handler functions.

TIMER INTERRUPTS

As well as catching interrupts from a digital input, you can also set periodic interrupts from a hardware timer running on the Pico. For example, we could change the LED blinking program to use a timer interrupt, so that it works without the need for calls to `sleep`. This frees up the eternal loop for other activities. You can find this program using the Pico's built-in LED, in 09_02_blink_timer.py.

```
from machine import Pin, Timer
from utime import sleep

led = Pin(25, Pin.OUT)

def tick(timer):
    global led
    led.toggle()

Timer().init(freq=2, mode=Timer.PERIODIC, callback=tick)

x = 0
while True:
    print(x)
    x += 1
    sleep(1.2)
```

Just to make the program a bit more interesting, the eternal loop counts up in the Shell—demonstrating that the timer is not adversely affecting anything else going on in the program.

The `tick` function is associated with a timed interrupt using the `init` method on `Timer`. The frequency of interrupt (times per second) is specified in the `freq` parameter. The mode of `PERIODIC` means the interrupt will keep repeating. You can also specify `ONE_SHOT`, if you only want the timed interrupt to happen once. You can find out more about the `Timer` class at https://docs.micropython.org/en/latest/library/machine.Timer.html.

It is a good idea to keep the code in an interrupt service routine as short and quick as possible, because, even though it may look like the Pico is doing two things at the same time, the counting is actually interrupted for a very short amount of time while the LED is being toggled in the `tick` function.

You will find another example using this OLED display to make a clock in Chapter 12.

Advanced Inputs and Outputs

HIGH SPEED PWM

This section is really only here because, when I was experimenting with my Pico, I was really impressed by its PWM capability.

The Raspberry Pi Pico is not unusual in providing a PWM output—all modern microcontrollers do this, but what is unusual is the resolution and speed of the PWM that the Pico can produce. If you are controlling the brightness of an LED, or the speed of a motor, then you do not need very high speed PWM, but, if you are producing audio signals, then high speed and high resolution make it much easier to produce high-quality analog output signals.

The Pico is capable of producing a 16 bit PWM output at a frequency of 10MHz or more. See Figure 9-2 where program 09_fast_pwm.py is being used to turn a GP16 on and off 2 million times a second!

```
from machine import Pin, PWM

out_pin = PWM(Pin(16))
out_pin.freq(10000000) # 10 MHz

out_pin.duty_u16(32768) # 50%
```

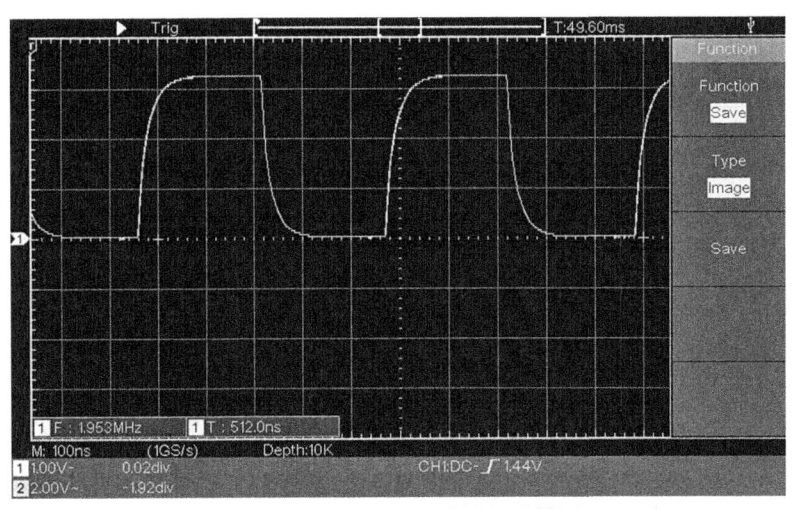

FIGURE 9-2 High speed PWM—2MHz, 50% duty cycle.

107

MULTICORE SUPPORT

If you come from the world of software development, you will be used to your programs being able to do more than one thing at a time—*multi-threading* as it is called. This relies on the operating system switching between processes, to give the appearance of multiple things hapening at the same time, or making use of multiple processors. Most microcontrollers don't use an operating system to schedule processes, and operate on a single processor (core). You can usually work around this limitation by using interrupts (as we saw earlier), but the Pico can genuinely do two things at once.

The Pico is a dual-core (two processor) device which means that, even without an operating system, it can assign activities to two different processes, allowing it to really do two things at once. Often, in an embedded system, two things at once is about right—one eternal loop can be working the user interface (waiting for button presses and displaying things) while the other controls something.

We can rewrite our interrupt example to make use of two cores—one core just counting, while the other core waits for a button press. You will find this example in 09_04_multicore.py.

```
from machine import Pin
from utime import sleep
import _thread

switch = Pin(10, Pin.IN, Pin.PULL_UP)

def core0():
    x = 0
    while True:
        x += 1
        print(x)
        sleep(1)

def core1():
    while True:
        if switch.value() == 0:
            print("button pressed")
```

Advanced Inputs and Outputs

```
        sleep(0.1)

_thread.start_new_thread(core1, ( ))
core0()
```

A good way to separate what each core will do, is to define a function into which that core's code will go. So, in this case, `core0` initialises a counter variable to 0 and then loops forever, adding 1 to x and printing it out.

The other core (`core1`) has a second eternal loop, that watches for a switch press and prints out a message when it happens.

The function core1 is started by calling the `start_new_thread` method on `_thread`. In the case of the Pico, `_thread` is just a way of identifying the second core (core 1).

The function `core0` is just started as a regular function call.

PROGRAMMABLE IO

The Pico can provide a wealth of features on most of its GPIO pins, including the following

- Normal digital inputs and outputs
- High speed PWM output on any pin
- Analog input on selected pins
- I2C and SPI bus interfaces on various pins
- UART (Serial interface) on various pins
- Interrupts on any pin

However, if it turns out that there is some interface you need that is missing from this list, then the Pico allows you to mix some low-level assembler code into your Python program which will allow one or more pins to implement that interface. We will see this in Chapter 11, where we interface a Pico to high speed, serial, addressable LEDs (often called NeoPixels). The stream of bits needed to do this is not one of the interfaces listed above but, none the less, we can add this feature to the Pico.

This means that the Pico community is continually developing and sharing new interfaces (called *state machines*) that we can make use of in our Python programs.

SUMMARY

In this chapter, we have discovered just how powerful the Pico is when it comes to inputs and outputs. Not only can we do things pretty fast, but we can also use cores to run two programs at the same time.

In Chapter 10, we will turn our attention to using sensors with the Pico, and try out some more electronics.

10

Sensors

In Chapter 8 we were mostly concerned with what we could call output devices such as LEDs and servomotors. In this chapter, we will look at sensors, that is, devices that measure some physical property, such as the position of a knob, temperature or light level.

Analog sensors turn physical values such as light level, temperature or position into a voltage that can be measured by an analog input of the Pico.

This is another chapter that makes use of the MonkMakes Electronics Kit 1 for Raspberry Pi Pico.

VARIABLE RESISTORS

Variable resistors are, as the name suggests, resistors whose resistance can be varied. In this case, the resistance is varied by changing the position of a knob. Figure 10-1 shows such a variable resistor. For historical reasons, relating to a common use of variable resistors to measure voltage, variable resistors are also often called *potentiometers* or just *pots*.

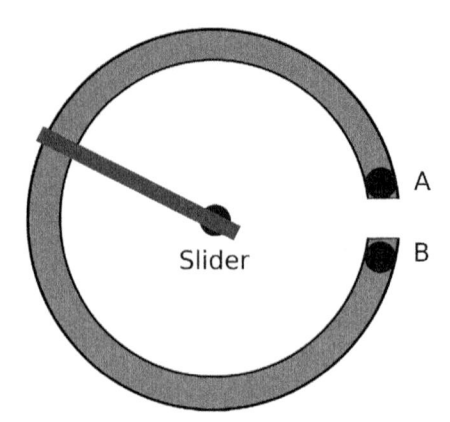

FIGURE 10-1 How a variable resistor works.

The resistor is made of a conductive track in an arc. The resistance between one end of the track and the other remains the same, but a central sliding contact (the slider) is moved by the variable resistor's knob, so that the resistance between the slider and one end of the variable resistor changes as you rotate the knob.

Resistance cannot be directly measured by an analog input. Analog inputs measure voltage, so we need a way of producing a variable voltage as the knob is turned. This is accomplished by the arrangement shown in Figure 10-2, called a voltage divider.

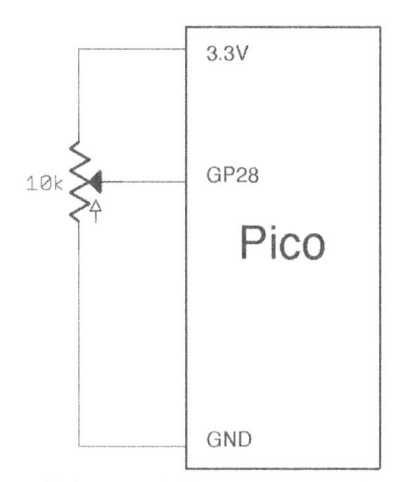

FIGURE 10-2 Using a variable resistor as a voltage divider.

Sensors

One end of the variable resistor is connected to GND and the other to 3.3V, and the slider of the variable resistor is connected to an analog input (GP 28). When the slider is at the GND end, the voltage at the slider will be 0V. When the slider is at the other end then it will be 3.3V and, at positions in-between, the voltage will be between 0 and 3.3V.

To test this out, wire up a variable resistor as shown in Figure 10-3.

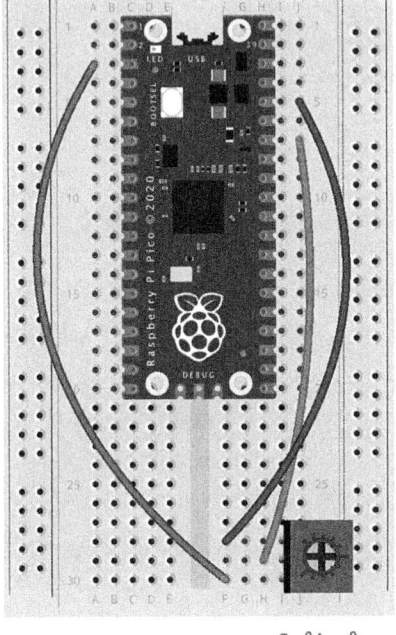

FIGURE 10-3 Wiring diagram for a variable resistor—pin 28 used.

To test the circuit, re-use the program 07_06_voltmeter.py from chapter 7. Turn the knob on the variable resistor to see how the voltage at the slider changes.

I have repeated the program here, as a reminder.

```
from machine import ADC, Pin
from utime import sleep
```

```
analog = ADC(28)

def volts_from_reading(reading):
    min_reading = 336
    max_reading = 65535
    reading_span = max_reading - min_reading
    volts_per_reading = 3.3 / reading_span
    volts = (reading - min_reading) * volts_per_reading
    return volts

while True:
    reading = analog.read_u16()
    print(volts_from_reading(reading))
    sleep(0.5)
```

TEMPERATURE

The Raspberry Pi Pico has a built-in temperature sensor, that is connected to analog channel 4. This pin is not available on the Pico's edge connector—it is only for use as an analog input for the built-in temperature sensor.

Open the program 10_01_thermometer.py and run it. You should see a stream of temperature readings appear. If you prefer your temperatures in Fahrenheit, then run the program 10_01_thermometer_f.py. Try putting your finger on the Pico's processor chip to warm it up—you should see the temperature rise.

```
from machine import Pin, ADC
from utime import sleep

temp_sensor = ADC(4)
points_per_volt = 3.3 / 65535

def read_temp_c():
    reading = temp_sensor.read_u16() * points_per_volt
    temp_c = 27 - (reading - 0.706)/0.001721
    return temp_c

while True:
    temp_c = read_temp_c()
```

```
print(temp_c)
sleep(0.5)
```

Rather confusingly, the analog inputs are referenced by their analog channel number (which for the temperature sensor is 4) rather than their GPIO number.

The math around calculating the temperature is taken from the Raspberry Pi Pico Python SDK documentation, which you can find here: https://datasheets.raspberrypi.org/pico/raspberry-pi-pico-python-sdk.pdf.

LIGHT

The MonkMakes Electronics Kit 1 for Raspberry Pi Pico includes a light sensor. This component is actually a phototransistor, that acts a bit like a variable resistor, whose resistance varies depending on the amount of light falling on it. Figure 10-4 shows the schematic diagram for connecting this sensor to an analog input.

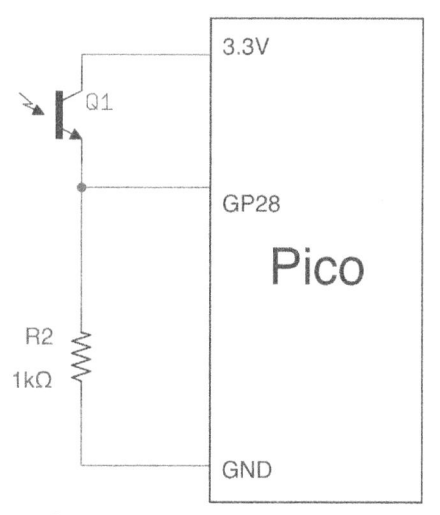

FIGURE 10-4 Schematic diagram for using a phototransistor to measure light intensity.

The 1kΩ resistor converts the change in current (as more light hits the phototransistor) into a voltage, that can be measured by the analog input. Wire up your breadboard as shown in Figure 10-5 to try this out. The longer lead of the phototransistor should be connected to row 7 (GP28).

FIGURE 10-5 Bereadboard layout for using a phototransistor to measure light intensity.

The program 10_02_light_meter.py will measure the light intensity and display it as a percentage from 0 (totally dark) to 100 (bright sunlight). Run it and try shading the phototransistor with your hand—the light level should change.

```
from machine import ADC, Pin
from utime import sleep
from math import sqrt

light_sensor = ADC(28)
dark_reading = 200
scale_factor = 2.5

def read_light():
```

```
    reading = light_sensor.read_u16()
    # print(reading)
    percent = int(sqrt(reading - dark_reading) / scale_factor)
    if percent < 0:
        percent = 0
    elif percent > 100:
        percent = 100
    return (percent)

while True:
    light_level = read_light()
    print(light_level)
    sleep(0.2)
```

Notice that we are importing `sqrt` (square root) from the `math` module. We will need this to scale our light reedings. The offset for readings in complete darkness and the sensitivity are controlled by the variables `dark_reading` and `scale_factor`. The value of `dark_reading` should be the raw analog reading when the phototransistor is in complete darkness—a value of 200 should be about right. Adjusting `scale_factor` controls the sensitivity so, if you are only going to be measuring the light indoors where it is dimmer than outdoors in the sunshine, then you might want to increase `scale_factor`.

After the raw analog reading is taken (`reading`), the `percent` is estimated, by taking the square root of the difference between the reading and the `dark_reading` and then multiplying by `scale_factor`. The square root is taken to give the scale a more natural curve, as perceived brightness is related to the square root of the light intensity.

The percentage reading is then clipped so that it is within the range 0 to 100.

SUMMARY

In this chapter we have looked at various different sensors that can be connected to the analog inputs of the Pico. In the next chapter we will turn our attention to displays and other output devices.

11

Displays

In this chapter, we will look at how to connect and use two common types of display with the Raspberry Pi Pico. The first is a common graphical OLED display. These displays are low-cost and readily available and, although a little small, are easy to use.

The second type of display uses addressable LEDs. These LEDs, often called *NeoPixels*, can be found in various arrangements, from long lengths of regularly spaced LEDs on tape, to rectangular arrays of LEDs.

OLED DISPLAYS

Low-cost OLED (Organic LED) displays, like the one shown in Figure 11-1, are widely available on the internet. Search for SSD1306 OLED, and look for displays that are 128x32 or 128x64 pixels in size. SSD1306 is the driver chip used by the OLED display.

These displays are bright and clear and are a stylish way of getting you Pico to display text or graphics.

Programming the Pico

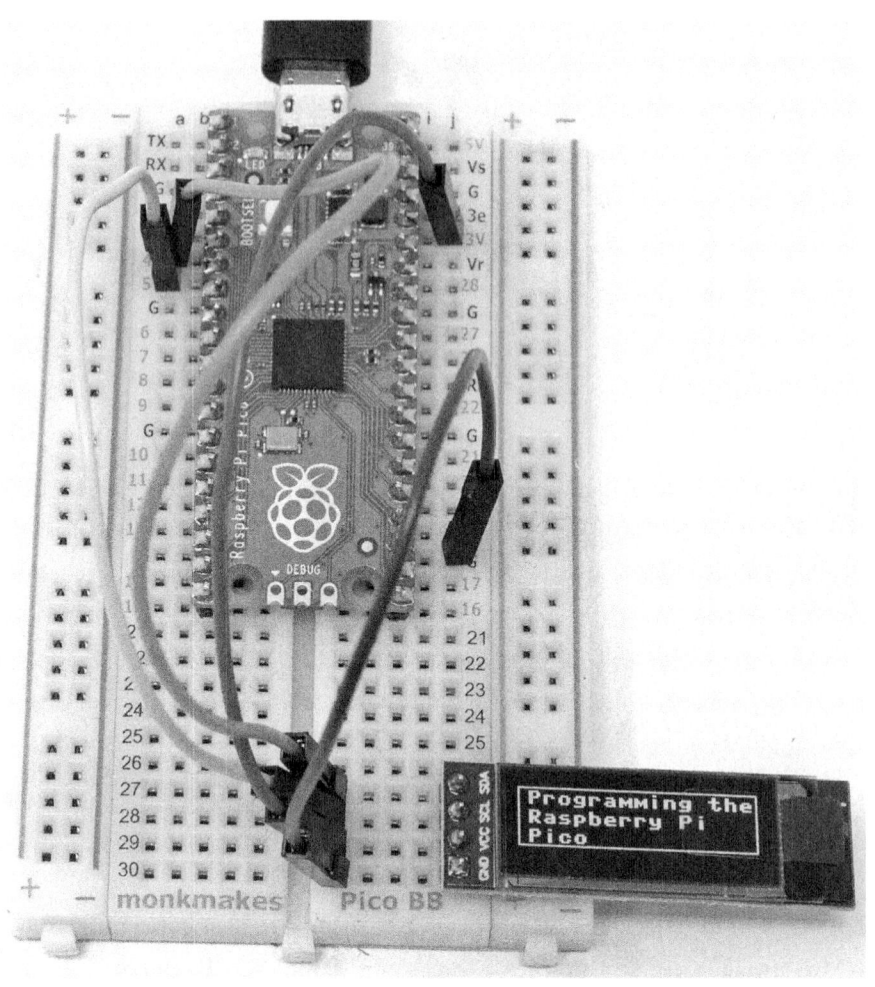

FIGURE 11-1 A low-cost 128 by 32 pixel OLD display.

Hardware

Figure 11-2 shows how to connect an OLED display to your Pico. The OLED display requires 3V power, and the SDA and SCL connections of the display are connected to GP 4 and 5 respectively, in what is known as a I2C interface.

Displays

FIGURE 11-2 Wiring an OLED display to a Pico.

Note that some OLED displays have the pins in a different order, so check what is written on the OLED display itself next to the pins.

The I2C interface usually requires two pull-up resistors to be used; however, the Pico's I2C interface uses the processor's internal pull-up resistors, so external resistors are not required.

Software

Using an OLED display like this requires the use of a Python module. So this is a good opportunity to see how we get a module onto a Pico. First, open the file called ssd1306.py in Thonny. You now need to get this file onto the Pico so, from the *File* menu, select *Save a Copy . . .* ; then chose *MicroPython Device* and save the file onto the Pico's file system with the name ssd1306.py (Figure 11-3).

Now that the module is on the Pico's file system, any program that we upload that needs to use the module will have access to it. So, open the program 11_01_oled.py and run it. Of course, if you want it to start automatically, you will have to save a copy, renaming it as main.py on the Pico.

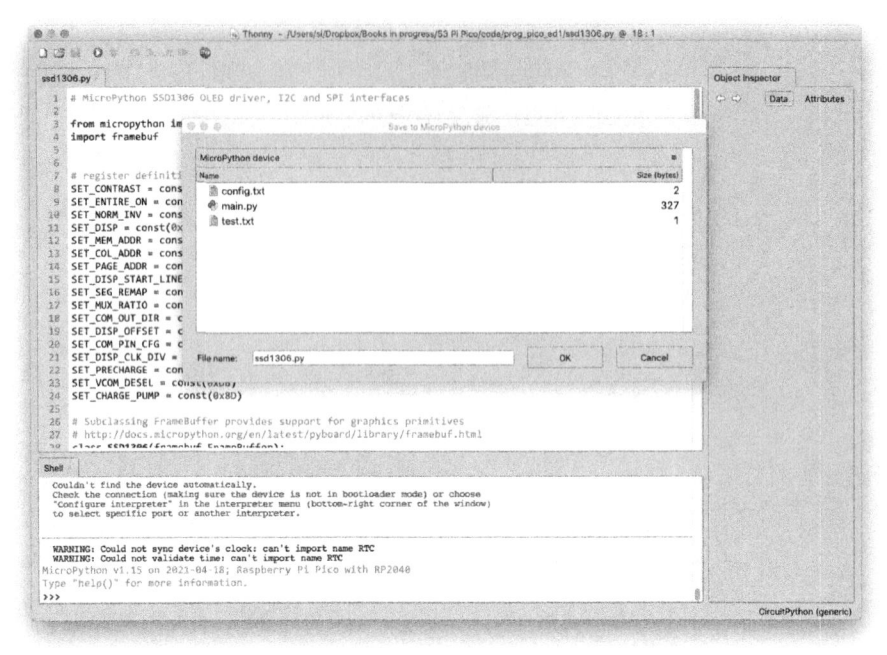

FIGURE 11-3 Saving a module onto the Pico.

Here's the code for the project.

```
from machine import Pin, I2C
from ssd1306 import SSD1306_I2C

i2c = I2C(0, sda=Pin(4), scl=Pin(5))
oled = SSD1306_I2C(128, 32, i2c)

oled.fill(0)
oled.rect(0, 0, 127, 31, 1)
oled.text("Programming the", 5, 2, 1)
oled.text("Raspberry Pi", 5, 12, 1)
oled.text("Pico", 5, 22, 1)

oled.show()
```

From the machine module, we need to import Pin and also I2C. I2C (pronounced I squared C) is a standard way of connecting displays and other modules to a microcontroller like the Pico. It uses two GPIO pins, and sends data back and forth at high speed.

A variable `i2c` is used to represent the I2C interface that we are going to use on the Pico. The Pico actually has two I2C channels and each of these channels can be allocated to different pair of pins. We are going to use the first I2C channel (0) on pins 4 and 5. We specify this by using the line

```
i2c = I2C(0, sda=Pin(4), scl=Pin(5))
```

The parameter `sda` refers to the I2C data connection, and `scl` to the clock connection.

The next line declares a variable `oled` to represent the display. The first two parameters are the display resolution (in this case 128 wide by 32 high); the final parameter links the `oled` display to the `i2c` interface.

Now that the display is all set up, we can display things. The display used here is monochrome (blue on black); other colors are available, and also multi-color oled displays. However, for this display, a value of 0 signifies a pixel is off and 1 that it is on. So, to clear the entire screen, we can fill it with 0 like this

```
oled.fill(0)
```

To draw a rectangle we need to specify the starting x and y coordinates (0, 0 is top left) and then the width and height of the rectangle in pixels (n this case 127 x 31). The final parameter to `oled.rect` is the color (in this case 1).

We can also display text, in which case the parameters are the text to display, followed by the x and y coordinates for the top left of the text, followed by the color (1).

Any of these drawing commands will not have any effect until we call `oled.show`. It us usual to prepare all of the things you want to display first and then use a single `oled.show` rather than do `oled.show` after every drawing commands.

AN OLED CLOCK

Back in Chapter 9 we used a timed interrupt which we can combine with our OLED display to create a crude clock.

This project uses the same breadboard layout as Figure 11-2 and you will find the program in 11_02_oled_clock.py.

When you run the program, the time will be displayed as 12:00:00. To set the correct time, change the line h, m, s = (12, 0, 0) to the current time before running the program—you'll have to be quick!

```
from machine import Pin, I2C, Timer
from ssd1306 import SSD1306_I2C

i2c = I2C(0, sda=Pin(4), scl=Pin(5))
oled = SSD1306_I2C(128, 32, i2c)

h, m, s = (12, 0, 0)

def show_time():
    oled.fill(0)
    time_str = "{:02d}:{:02d}:{:02d}".format(h, m, s)
    print(time_str)
    oled.text(time_str, 0, 0, 1)
    oled.show()

def tick(timer):
    global h, m, s
    s += 1
    if s == 60:
        s = 0
        m += 1
        if m == 60:
            m = 0
            h += 1
            if h == 24:
                h = 0
    show_time()

Timer().init(freq=1, mode=Timer.PERIODIC, callback=tick)
```

Every second (`freq=1`) the function `tick` is called. This uses a cascading sequence of `if` statements to update the time held in three variables `h`, `m` and `s` (hours, minutes and seconds). Once the seconds count `s` reaches 60, it is reset to 0 and 1 added to the minutes count `m`. A similar process happens for the minutes and hours. At the end of the `tick` function, the `show_time` function is called to update the OLED display.

MicroPython's string formatting feature is used to insert `h`, `m` and `s` into a formatting string (`{:02d}:{:02d}:{:02d}`) that substitutes the three numbers into place holders within the string, enclosed in `{}`. Using `{:02d}` ensures that each number is exactly 2 digits long, so that the display does not look strange for times with single digit minutes and seconds. You can find out more about the format method at https://www.w3schools.com/python/ref_string_format.asp.

NEOPIXEL DISPLAYS

NeoPixel has become a popular term for displays using chains of addressable LEDs—LED chips that include red, green and blue LEDs, as well as control logic to set the color. Such displays are available in all sorts of shapes and sizes, from long tapes with adhesive backing, or rings like the one shown in Figure 11-4, to rectangular arrays of LEDs arranged as a screen with very large pixels.

Hardware

Adafruit have a nice range of NeoPixel displays, and you can also find similar products at a lower cost (made in China). The key thing to look for is *WS2812*, which is the name of the NeoPixel addressable LED chip used for the displays.

Theses displays require a logic level that is close to the supply voltage; by preference this would be 5V, but although the Pico can supply 5V at a reasonable current, its logic level is 3.3V. So, if you try to supply the display with 5V, but use the 3V data logic of the Pico, the results are unpredictable. Hence, it is better to use a 3V supply. This means that there is much less current available to

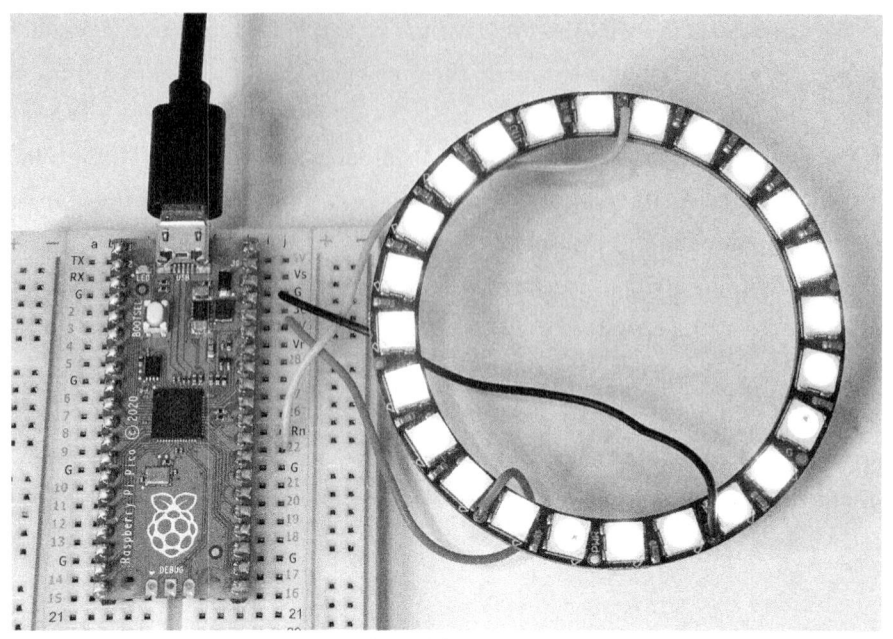

FIGURE 11-4 An Adafruit NeoPixel display.

power the LEDs, so you will need to limit either the brightness of the LEDs (in software) and/or the number of LEDs. Otherwise the display will take too much current, the voltage will then drop, and the display and/or Pico will not operate reliably. So, if things start to misbehave, that's probably what is happening and you need to reduce the brightness of the LEDs or use a separate high current power supply to provide power to the NeoPixels.

However many LEDs you have in your display, you will have just three connections to be made—power (3V and GND) and data (which is connected to a GPIO pin). In the code example in the next section, it is assumed that the data connection is connected to pin GP22.

You can make the connections using breadboard like Figure 11-4, where wires are soldered to the display and then connected to the breadboard. Alternatively, you may find that you can connect to the display using jumper wires.

Software

The software for this type of display is particularly interesting, because it makes use of the Pico's ability to define little programs (called *state machines*) that actually run on the GPIO pins them-selves, not requiring any effort from the Pico's processors.

Since the program is quite long, we will deal with it in parts—but you should follow the description, referring to the file 12_02_neopixel.py.

After the imports, and the variable NUM_LEDS that you need to set to the number of LEDs in your display, you will see a section of code that sets up GP22 to feed the display with the high speed serial data it requires.

```
@rp2.asm_pio(sideset_init=rp2.PIO.OUT_LOW,
        out_shiftdir=rp2.PIO.SHIFT_LEFT,
        autopull=True,
        pull_thresh=24)

def ws2812():
    T1 = 2
    T2 = 5
    T3 = 3
    wrap_target()
    label("bitloop")
    out(x, 1)               .side(0)    [T3 - 1]
    jmp(not_x, "do_zero")   .side(1)    [T1 - 1]
    jmp("bitloop")          .side(1)    [T2 - 1]
    label("do_zero")
    nop()                   .side(0)    [T2 - 1]
    wrap()

sm = rp2.StateMachine(0, ws2812, freq=8_000_000,
    sideset_base=Pin(22))
sm.active(1) # Start the StateMachine
pixels = array.array("I", [0 for _ in range(NUM_LEDS)])
```

This code does not look much like any Python we have seen so far. That's because it's actually using the Python syntax to define code in a type of low-level code called *assembly language*. Fortunately that doesn't matter too much as we can just take this as a *black box* that we can use in our programs whenever we want to use Neopixels. So if the next two paragraphs seem obscure—don't worry!

The basic idea is that the Pico's processor has some dedicated hardware for these state machines that greatly facilitates sending out high-speed serial data on one or more GPIO pins.

If you have programmed in assembler, then you should be able to decode some of what is going on in the function ws2812. This function contains the mini-program or state machine that will be run by this specialized hardware. For example, the call to label("bitloop") marks a point where we can jump to in this mini program. Then later on we have a call to jmp("bitloop"), which will cause a jump back to that point in the state machine.

The list pixels is a list of 24-bit color values, one element for each LED. So, to set the color of a particular LED, you just need to put a new value in the list element for that pixel. To make the use of this array easier, there are three functions defined that you would probably want to copy to any Neopixel projects you make.

```
def set_led(led, red, green, blue):
    pixels[led] = (red << 16) + (green << 8) + blue

def show():
    sm.put(pixels, 8)

def clear():
    for i in range(NUM_LEDS):
        set_led(i, 0, 0, 0)
    show()
```

The function set_led takes as its parameters the number of the LED whose color you want to set (starting at 0) followed by the red, green and blue components of the color, each as a number between 0 and 255. Where a color value of 0 means off and 255 maximum brightness. However, as we discussed earlier, we are on a current budged here, so its worth keeping these values fairly low. A value under 50 will still make the LED quite bright, without using too much current.

Displays

To combine the three separate color values into one big number, we have to use something called *bit shifting* (see Figure 11-5).

Each of the smaller color values between 0 and 255 is represented in binary (1s and 0s). To represent a number between 0 and 255, you need 8 binary digits (bits). To put these three 8 bit numbers into a single 24 bit number, we need to multiply the red part by 65536 (16 bits) to push it to the left and make room for the green and blue parts. The green part must be multiplied by 256 (8 bits) to shift that to the left and then finally, the blue part is added into the rightmost 8 bits.

Although we could literally multiply the red value by 65536 and the green value by 256, it is more efficient for the processor to use << which moves the bits to the left by the value specified after the << symbol.

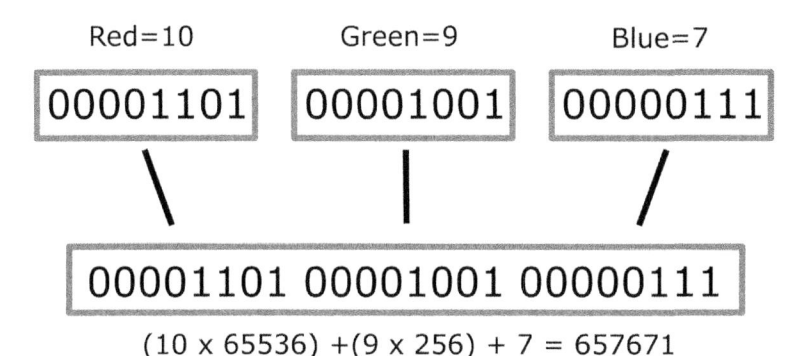

```
    Red=10          Green=9          Blue=7
  00001101         00001001         00000111
         \             |            /
         00001101 00001001 00000111
         (10 x 65536) +(9 x 256) + 7 = 657671
```

FIGURE 11-5   Bit-shifting three 8 bit numbers into one 24 bit number.

The show function works rather like the show function in the SSD1306 module from the previous section. That is, nothing will change on the display until after this is called.

The clear function will blank the display, by setting every element of the array to 0 and then calling show.

The rest of the code is essentially just there to show the display in action.

```
def randomize():
    clear()
    for i in range(NUM_LEDS):
        set_led(i, randint(0, 50), randint(0, 50),
            randint(0, 50))
        show()
        time.sleep(0.1)

clear()

print("Enter the LED's number to turn it on")
print("or c-clear r-randomize")
while True:
    led_str = input("command: ")
    if (led_str == 'c'):
        clear()
    elif (led_str == 'r'):
        randomize()
    else:
        led = int(led_str)
        set_led(led, 50, 50, 50) # white
        show()
```

The randomise function sets each of the LEDs to a randomly chosen color. This program is interactive, and the eternal loop asks you to enter a command (which can either be the number of an LED to turn on, the command c to clear the display, or the command r to set the LEDs to random colors).

SUMMARY

In this chapter we have used two very different types of display. There are, of course, many other types of display available, many of them using an I2C interface. If you have hardware that you want to connect to your Pico, search to see if someone has written a module for it—this is a lot easier than trying to write your own. Also, remember that CircuitPython has many such Python modules available, and your project may be more easily implemented in CircuitPython if no module for your hardware is available in MicroPython.

12
Other Programming Languages

MicroPython is the easiest and best place to start with your Pico. However, now that you have mastered the basics of MicroPython, you might like to experiment with some of the other programming options.

In this chapter, we first look at another flavour of Python (called *CircuitPython*) and then switch languages entirely to look at the use of C++ with the Pico, using the Arduino environment.

Experienced embedded programmers may eschew the Arduino C++ environment in favour of mbed or other more professional C++ ARM environments. I don't discuss these because, if you are that advanced, you probably don't need this book!

CIRCUITPYTHON

CircuitPython has the same roots as MicroPython—in programming terms, CircuitPython is a *fork* of MicroPython. That is, the open-source MicroPython code was copied, and then taken in a slightly different direction by the American company Adafruit. The software is still open-source, but Adafruit have taken the original MicroPython and added a whole load of code to it. In particular, they have added module support for a host of hardware add-ons that can be attached to the Pico.

This effort is not entirely altruistic, as Adafruit sell those hardware products for which they have written modules, but, most of the code is compatible both with Adafruit products and other hardware from different manufacturers (including low-cost imported modules from China).

Adafruit also manufacture a range of boards such as the Feather RP2040 (Figure 12-1) that use the Pico's processor on a board of their own design (a Pico-compatible if you like).

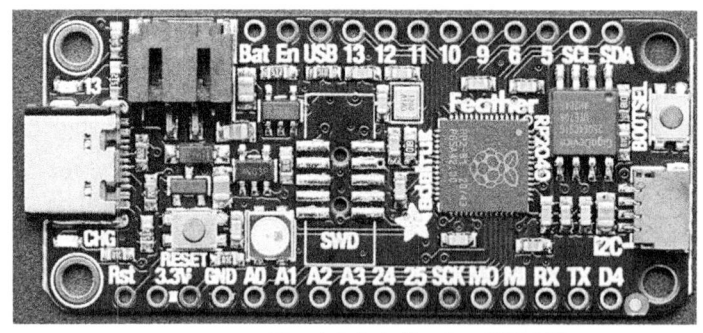

FIGURE 12-1 The Adafruit Feather RP2040.

Although Thonny does offer some support for CircuitPython, Adafruit recommends the use of the Mu editor. Mu is similar in many ways to Thonny, being a light-weight and easy to use editor.

We explain how to install all of these below.

CircuitPython vs MicroPython

When planning this book, the decision as to whether to base it on MicroPython or CircuitPython was quite evenly balanced. On the one hand, MicroPython is the environment most supported by the Raspberry Pi Foundation, and therefore the most likely starting point for people. On the other hand, when it comes to attaching peripherals such as displays and advanced sensors to the Pico,

Other Programming Languages

then CircuitPython has the advantage (at the time of writing). In the end I took the view that it doesn't actually matter which you learn first as, if you know how to use one, its pretty easy to get up to speed on the other.

Installing CircuitPython

The first time you ran Thonny with your Pico attached to your computer, you were asked if you wanted to install the firmware for MicroPython. This process then copied a UF2 ROM image onto the Pico, installing MicroPython. Installing CircuitPython is a similar process, except you have to install it manually. To do this, first put your Pico into *boot* mode, by unplugging it and then holding down the BOOTSEL button (Figure 12-2) for 3 seconds after you plug it back in again.

FIGURE 12-2 The BOOTSEL button.

When you do that, the Pico enters a special mode for updating its firmware and will appear in your file system with a new name of RPI-RP2. You now need to download the UF2 file for CircuitPython, to replace MicroPython. Note that this process is reversible, and you can always go back to MicroPython using Thonny to reinstall MicroPython.

To get the UF2 file, go to https://circuitpython.org/board/raspberry_pi_pico/ (Figure 12-3) and click on DOWNLOAD. Depending on your browser and operating system, you may be able to save the file directly onto the RPI-RP2 location. If not, you will have to save it somewhere else, and then copy the file onto RPI-RP2.

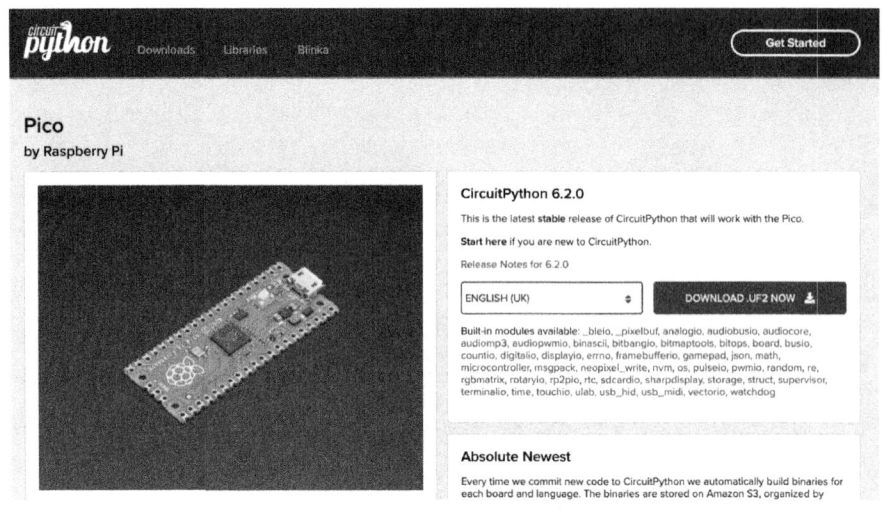

FIGURE 12-3 Downloading the UF2 CircuitPython image.

Once the copying is complete, your Pico should now appear in your file system with the name CIRCUITPY—and be ready to use with CircuitPython.

Other Programming Languages

Installing Mu

Follow the instructions at https://codewith.mu/en/download to download the version of Mu for your operating system. The first time that you run Mu, you need to select the Mode it operates in (Figure 12-4)—select the mode CircuitPython.

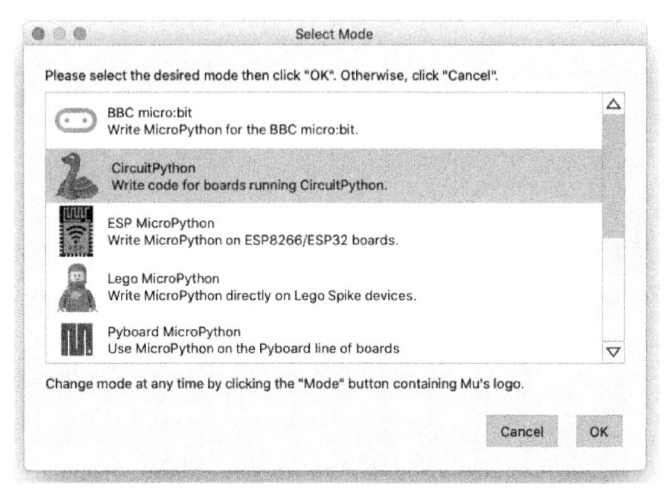

FIGURE 12-4 Selecting the CircuitPython mode in Mu.

Blinking Revisited

Let's start with the basics and write our built-in LED blinking program in CircuitPython and put it alongside the MicroPython version. You will find this code in the file 12_01_circuit_python_blink.py, but running it on the Pico requires you to write it to a file called main.py on the pico itself. So, having opened the file, click on the New button to create a new empty file, and then copy and paste the contents of 12_01_circuit_python_blink.py into it (Figure 12-5). Then save the new file to main.py on the Pico (CIRCUITPY).

Programming the Pico

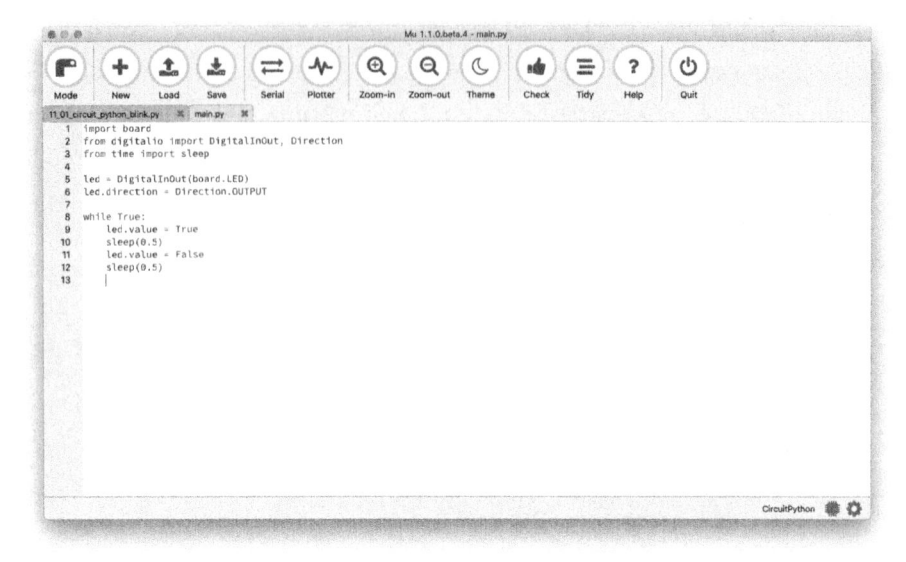

FIGURE 12-5 Editing main.py on the Pico.

You are now editing the program file directly on the Pico so remember that, if you need to save your work onto something more permanent than the Pico you will need to copy the file onto your computer's disk.

Table 12-1 shows the MicroPython and CircuitPython versions of Blink side by side.

TABLE 12-1 Blink in CircuitPython and MicroPython

CircuitPython	MicroPython

`import board`
`from digitalio import`
`   DigitalInOut, Direction`
`from time import sleep`
 	
`led = DigitalInOut(board.LED)`
`led.direction = Direction.OUTPUT`
 	
`while True:`
`    led.value = True`
`    sleep(0.5)`
`    led.value = False`
`    sleep(0.5)`
	`from machine import Pin`
	`from utime import sleep`
	
	`led = Pin(25, Pin.OUT)`
	
	`while True:`
	`    led.on()`
	`    sleep(0.5) # pause`
	`    led.off()`
	`    sleep(0.5)`

In CircuitPython, things to do with the Pico's hardware, such as the pins and built-in LED are imported from the board module rather than the machine module of MicroPython. In CircuitPython, the more unwieldy DigitalInOut is used in place of MicroPython's Pin.

Setting the direction of the pin (input or output) is a separate step to defining the pin to use with CircuitPython, and the built-in LED is referenced as just board.LED, rather than having to specify the pin number of 25.

If you want to learn more about CircuitPython, then you will find many excellent resources available at the CircuitPython website (https://circuitpython.org/) and at Adafruit (https://learn.adafruit.com/welcome-to-circuitpython/what-is-circuitpython).

ARDUINO C/C++

Probably the biggest single tool making microcontrollers accessible to makers and hobbyists was the Arduino board and its associated integrated development environment (IDE). Arduino wrapped up microcontrollers into an easy-to-use, open-source package. To this day, the Arduino IDE's open architecture has allowed it to be expanded to support almost any board imaginable, including the Raspberry Pi Pico.

At the time of writing, there are two ways of using a Pico with the Arduino IDE—the official Arduino version, and the unofficial plug-in core created by Earle F. Philhower, III. Of these, at the time of writing, the unofficial version is the better. You can find out more about it here, including instructions on installation: https://github.com/earlephilhower/arduino-pico.

Once installed, start the Arduino IDE and then load the Blink example sketch from the Files->Examples->01. Basics menu (Figure 12-6).

Now, you need to tell Arduino which board you are using, so select Raspberry Pi Pico from the Tools->Boards menu. You can accept the default options for the board (Figure 12-7).

Programming the Pico

FIGURE 12-6 Blink in the Arduino IDE.

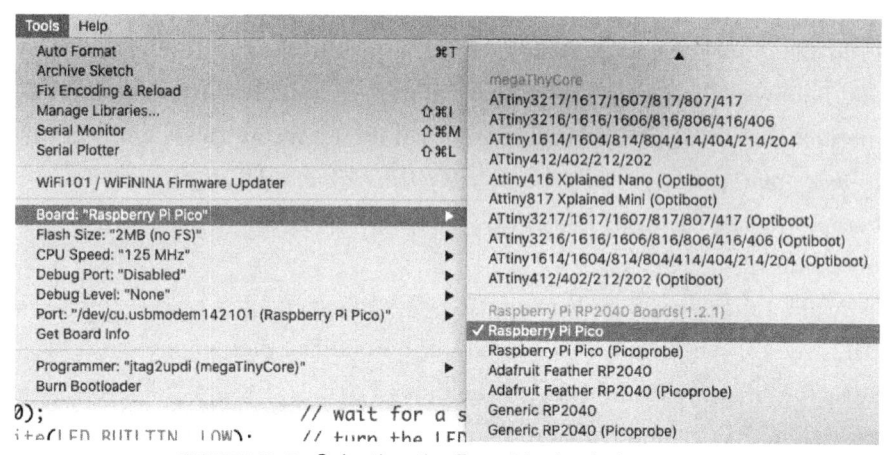

FIGURE 12-7 Selecting the Board in the Arduino IDE.

Click on the Upload button and the blink program (or sketch as they are called in Arduino) will upload onto the Pico and start running. Lets look at the Arduino code for blinking.

```
void setup() {
  pinMode(LED_BUILTIN, OUTPUT);
}
```

138

```
void loop() {
  digitalWrite(LED_BUILTIN, HIGH);
  delay(1000); // milliseconds
  digitalWrite(LED_BUILTIN, LOW);
  delay(1000);
}
```

In C and C++. comments are marked using // rather than the # of Python. So that we can see the code properly, I've deleted the comments from the Blink example.

Arduino programs must define two functions called `setup` and `loop`. The function `setup` will automatically be run just once when the Pico powers up or is reset. In this case, we are setting the pin mode of LED_BUILTIN (the built-in LED on pin 25) to be an output.

The `loop` function is the Arduino equivalent of the eternal loop and any code you put in here will be run over and over again. In this case, we use the `digitalWrite` and `delay` functions to blink the LED.

Most code samples written for Arduino should work with the Pico. The Pico version even supports the dual-core feature of the Pico's processor, which means that you can take advantage of the vast array of community Arduino libraries that people have developed.

SUMMARY

MicroPython is probably the best way to get started programming your Pico; however, the alternative CircuitPython provides a richer array of interfaces to hardware than CircuitPython. If you prefer C, then the easiest way into Pico programming is to use the Arduino IDE.

No doubt, by the time you are reading this book, other options will also be available.

Appendix

RASPBERRY PI PICO PINOUT

Appendix

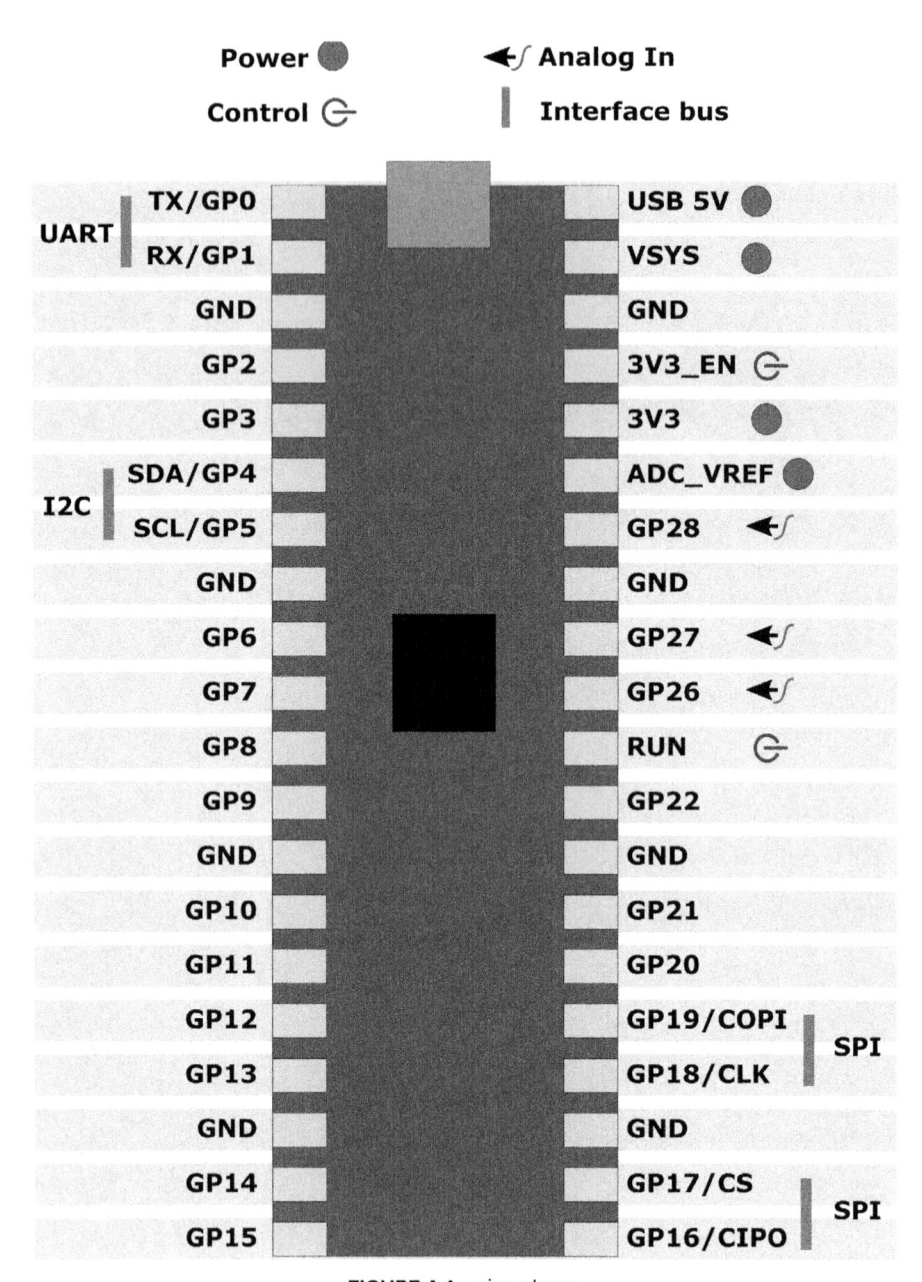

FIGURE A-1 pinout.png

Index

Symbols
< (less than), 32
> (greater than), 32
<= (less than or equal to), 32
>= (greater than or equal to), 32
/ (divide)
* (multiply)

A
ADC class, 81
Analog output, 77
Analog input, 80
Arduino, ix, 137

B
Bit shifting, 129
Bootsel (boot select) button, 133
Bouncing (switch contacts), 103
Breadboard, Solderless, 86
Buzzer, 91

C
C++, 137
Conditional code, 31
CircuitPython, 131
  installing, 132
  vs. MicroPython, 132

Classes, definition, 63
Clock project, 124
Code, downloading examples, 27
Components (electronic), 90
Configuration files, 71
Cores (dual), 108

D
Data busses on the Pico, 6
Dictionaries, 50
  accessing elements, 50
  in Python, 50
Dice, 40
Digital Inputs, 74
Digital outputs, 73
Display, 119
  OLED module, 119
  Neopixels, 125
Downloading, companion code for book, 27

E
else command, 31
Electronic components, 90
Embedded computing, 2
Eternal loop, 25
Exceptions, 70

Index

F
File, 68
  closing, 69
  counting example, 71
  opening, 69
  reading, 70
  writing, 69
find method, 66
for loop, 29
format method, 66
Fritzing, xiii
Functions, 35
  definition, 35
  parameters, 36

G
GitHub, dowloading code from, 27
GPIO (general Purpose Input Output) pins, 4, 70

H
Hello World program in Python, 23

I
I2C bus, 6, 120
if command, 31
Inputs
  digital, 74
  analog, 80
Interrupts, 103
  from digital input, 103
  timer driven, 105
Indentation, in Python, 26
ISR (Interrupt Service Routine), 104

J
Jumper wire, 91

L
LED electronic component, 91
  built-in, 15
  connecting to Pico, 92
  red, 91
  RGB, 91, 94
led.off method, 16
led.on method, 16
led.toggle method, 74
led.value method, 74
len function, 64
Light, sensing, 115
Linux, Thonny installation on, 10
Lists , 45
  in Python, 45
  of numbers, 48
Looping
  for, 29
  while, 25
  eternal, 25
lower method, 64

M
Mac OS, Thonny installation on, 10
main.py, 24
math module, 62
Microcontroller, definition, 1
MicroPython, 19
Modules, 59
  definition, 59
  importing, 59
Morse Code translator, 52
Mu Python editor, 132
Multi-core support, 108

N
Named parameters, 41
Neopixels, 125
Numbers in Python, 19

Index

O
Object-orientation, 63
OLED, 119
Outputs
    digital, 73
    analog, 77

P
Paperclip, as a switch, 76
Parameters, to functions, 36
Parameters, named, 41
pass command, 60
Phototransistor, 91
Pico
    compatible boards (third party), 7
    connecting to a computer, 14
    pinout, 5
    tour of, 3
Pin class, 75
Pins, soldering onto a Pico, 88
Pot (potentiometer or variable resistor), 91
Power supply of Pico, 3
Programming, definition, 7
Programmable Input/Output, 109, 127
    Pull-up resistor, 75
Push switch, 91
PWM (Pulse Width Modulation), 77, 107

R
random, Python module, 62, 20, 40
Raspberry Pi, other models, ix
replace method, 67
Resistor, 91
return command, 39
Running programs automatically, 24

S
Shell, the Python, 15
Sensors, 111
    definition, 111
    light, 115
    temperature, 114
    variable resistor, 110
Servomotor, 91, 97
sleep method, 26
Soldering pins onto a Pico, 88
Solderless breadboard, 85
S.O.S. example, 28
SPI bus, 6
split method, 67
Strings, in Python, 22, 48
strip method, 68
Substrings, 49
Switch, 76

T
Thonny
    getting started, 11
    editor modes, 12
    installation, 9
Temperature, sensing, 114
Timer interrupts, 105

U
UART (Universal Asynchronous Receiver Transmitter) on the Pico, 6
USB lead, 15
utime module, 60

V
Variable in Python, 20
Variable resistor, 91, 110
Voltage regulator, 3

W
while loop, 25
Windows, Thonny installation on, 10

About the Author

Simon Monk has written over twenty titles on hobby electronics, and sold over 700,000 books that have been translated into ten different languages. He maintains that the best time to write a book about something is just after you have learned about it, as this is the sweet spot between knowing the subject and remembering the pitfalls.

You can follow him on Twitter, where he is @simonmonk2, and find out more about his books at http://simonmonk.org.

Printed in Great Britain
by Amazon